NO good

82-0153

WITHDRAWN
University of
Illinois Library
at Urbana-Champaign

VLSI 81
Very Large Scale Integration

Proceedings of the first International Conference on Very Large Scale Integration held at the University of Edinburgh from 18–21 August 1981, organised by the University of Edinburgh Departments of Computer Science and Electrical Engineering and the Wolfson Microelectronics Institute, with the assistance of CEP Consultants Ltd, 26 Albany Street, Edinburgh EH1 3QH

VLSI 81

Very Large Scale Integration

Edited by

John P. Gray

*Department of Computer Science
University of Edinburgh*

1981

ACADEMIC PRESS
A Subsidiary of Harcourt Brace Jovanovich, Publishers
London New York Toronto Sydney San Francisco

ACADEMIC PRESS INC. (LONDON) LTD.
24/28 Oval Road,
London NW1

United States Edition published by
ACADEMIC PRESS INC.
111 Fifth Avenue
New York, New York 10003

Copyright © 1981 by
ACADEMIC PRESS INC. (LONDON) LTD.

All Rights Reserved
No part of this book may be reproduced in any form by photostat, microfilm, or any other means, without written permission from the publishers

British Library Cataloguing in Publication Data
VLSI 81.
   1. Integrated circuits — Large scale integration — Congresses
  I. Gray, John P.
  621.381'73   TK874

ISBN 0-12-296860-3

LCCCN 81-68035

Printed in Great Britain by
Whitstable Litho Ltd., Whitstable, Kent

PREFACE

VLSI 81 is the first European Conference dedicated to all the subjects involved in the exploitation of silicon as a systems implementation medium. It has only recently become apparent, due to the pioneering work of Mead, that this emerging area of research embraces a very wide range of disciplines from device physics to branches of discrete mathematics.

One of the goals of the Programme Committee was to reflect this diversity by putting together a broad programme. Interestingly, many of the papers also reflect this diversity by bridging a number of apparently disparate subjects. Special emphasis has also been given to the more theoretical aspects of the subject. This is to give increased visibility to the areas which hold, the Committee believe, the more challenging problems, and more fundamental results, for progress in this subject.

John P. Gray
Chairman
Programme Committee

PROGRAMME COMMITTEE

J P Gray, Chairman, University of Edinburgh, UK
W Laing, Secretary, University of Edinburgh, UK
P Antognetti, University of Genoa, Italy
J Borel, EFCIS Grenoble, France
B G Bosch, Ruhr-Universität Bochum, FRG
A N Broers, IBM, USA
I Buchanan, University of Edinburgh, UK
D D Buss, Texas Instruments, USA
J Clark, Stanford University, USA
H de Man, Catholic University of Leuven, Belgium
D Eglin, International Computers Ltd, UK
W Heller, IBM, USA
S Kelly, General Instruments Ltd, UK
D Kinniment, Newcastle University, UK
F M Klaasen, Philips Research Laboratories, The Netherlands
D W Lewin, Brunel University, UK
J Mavor, University of Edinburgh, UK
D McCaughan, GEC Hirst Research Centre, UK
A D Milne, University of Edinburgh, UK
R Milne, Inmos Ltd, UK
R Milner, University of Edinburgh, UK
J P Mucha, University of Hanover, FRG
J C Mudge, CSIRO, Australia
M Newell, Xerox PARC, USA
D O Pederson, University of California, Berkeley, USA
F Preparata, University of Urbana, USA
D J Rees, University of Edinburgh, UK
M Rem, Eindhoven University of Technology, The Netherlands
J G L Rhodes, Pye TMC Ltd, UK
N Weste, Bell Laboratories, USA

ORGANISING COMMITTEE

From the University of Edinburgh Departments of Computer Science, Electrical Engineering and the Wolfson Microelectronics Institute:

S Michaelson (Chairman)
G Plotkin, D J Rees (Joint Secretaries)
J P Gray, W Laing, J Mavor, A D Milne
P D Schofield, J B Tansley

ADDITIONAL REFEREES

B Ackland, N F Benschop, G Brebner
M R Hannah, R P Kramer, P Rashidi
L Smith, L Valiant, M C Van Lier
R Vervoordeeldink, R Wynhoven

CO-SPONSORS

British Computer Society
European Association for Theoretical Computer Science
Institution of Electrical Engineers
Institute of Electrical and Electronics Engineers
    (Region 8)
Institute of Physics

ACKNOWLEDGEMENT

This International Conference is organised with the support of the following:

Burroughs Machines Ltd
Compeda Ltd
Hewlett Packard Ltd
IBM (UK) Ltd
Inmos Ltd
Prestwick Circuits Ltd
Scottish Development Agency
Standard Telecommunication Laboratories Ltd
Plessey-UK Ltd

AUTHORS

Ackland, B	117	Lyon, R F	131
Ahmed, H M	43	Marques, J A	53
Banatre, J-P	141	Mayle, N	183
Barton, E E	25	Mead, C A	3
Batali, J	183	Mikhail, W F	301
Blahut, D E	35	Miller, G L	289
Bryant, R E	329	Molzen, W W	95
Cardelli, L	173	Monier, L	269
Chang, H	95	Morf, M	43
Chazelle, B	269	Mosteller, R C	163
Colbry, B W	35	Mudge, J C	205
Collins, B	107	Mueller-Glaser, K D	319
Courtois, B	341	Myers, D J	151
Denyer, P B	151	Nair, R	257
Donath, W E	301	Parker, A C	357
Foster, M J	75	Plotkin, G	173
Frison, P	141	Quinton, P	141
Gordon, M	85	Rem, M	65
Gray, A	107	Roth, J P	351
Hafer, L J	357	Rupp, C R	227
Harrison, M	35	Séquin, C H	13
Hong, S J	257	Shapiro, E	257
Hwang, J P	95	Shrobe, H	183
Kinniment, D J	193	Smith, K F	247
Krambeck, R H	35	Snyder, L	237
Kuhn, R H	279	So, H C	35
Kung, H T	75	Soukup, J	35
		Sussman, G	183
Larkin, M W	313	Weise, D	183
Law, H F S	35	Weste, N	117
Leighton, F T	289	Whitney, T	217
Lerach, L	319	Wu, J C	95

CONTENTS

Preface	v
Programme Committee, Organising Committee	vii
Additional Referees, Co-sponsors, Acknowledgement	viii
Authors	ix

SESSION 1

VLSI and Technological Innovation *C A Mead*	3
Generalized IC Layout Rules and Layout Representations *C H Séquin*	13
A Non-Metric Design Methodology for VLSI *E E Barton*	25

Top Down Design of a One Chip 32-Bit CPU
  *R H Krambeck, D E Blahut, H F S Law, B W Colbry
  H C So, M Harrison and J Soukup*
	35

Synthesis and Control of Signal Processing
  Architectures Based on Rotations
  H M Ahmed and M Morf
	43

Mosaic: A Modular Architecture for VLSI System Circuits *J A Marques*	53

SESSION 2

The VLSI Challenge: Complexity Bridling *M Rem*	65

Recognize Regular Languages with Programmable
  Building-Blocks
  M J Foster and H T Kung
	75

A Very Simple Model of Sequential Behavior of nMOS *M Gordon*	85
Magnetic-Bubble VLSI Integrated Systems *H Chang, W W Molzen, J P Hwang and J C Wu*	95

SESSION 3

The Inmos Hardware Description Language and Interactive Simulator
B Collins and A Gray   107

A Pragmatic Approach to Topological Symbolic IC Design
N Weste and B Ackland   117

A Bit-Serial VLSI Architectural Methodology for Signal Processing
R F Lyon   131

A Network for the Detection of Words in Continuous Speech
J-P Banatre, P Frison and P Quinton   141

Carry-Save Arrays for VLSI Signal Processing
P B Denyer and D J Myers   151

SESSION 4

REST - A Leaf Cell Design System
R C Mosteller   163

An Algebraic Approach to VLSI Design
L Cardelli and G Plotkin   173

The DPL/Daedalus Design Environment
J Batali, N Mayle, H Shrobe, G Sussman and D Weise   183

Regular Programmable Control Structures
D J Kinniment   193

SESSION 5

VLSI Chip Design at the Crossroads
J C Mudge   205

A Hierarchical Design Analysis Front End
T Whitney   217

Components of a Silicon Compiler System
C R Rupp   227

Overview of the CHiP Computer
L Snyder   237

Implementation of SLA's in NMOS Technology
K F Smith   247

A Physical Design Machine
S J Hong, R Nair and E Shapiro   257

SESSION 6

Optimality in VLSI *B Chazelle and L Monier*	269
Chip Bandwidth Bounds by Logic-Memory Tradeoffs *R H Kuhn*	279
Optimal Layouts for Small Shuffle-Exchange Graphs *F T Leighton and G L Miller*	289
Wiring Space Estimation for Rectangular Gate Arrays *W E Donath and W F Mikhail*	301

SESSION 7

Impact of Technology on the Development of VLSI *M W Larkin*	313
A General Cell Approach for Special Purpose VLSI-Chips *K D Mueller-Glaser and L Lerach*	319
A Switch-Level Model of MOS Logic Circuits *R E Bryant*	329
Failure Mechanisms, Fault Hypotheses and Analytical Testing of LSI-NMOS (HMOS) Circuits *B Courtois*	341
Automatic Synthesis, Verification and Testing *J P Roth*	351
Automating the Design of Testable Hardware *A C Parker and L J Hafer*	357

SESSION 1

VLSI AND TECHNOLOGICAL INNOVATIONS

Carver A. Mead

*California Institute of Technology
Pasadena, CA 91125, USA*

Rather than innovation in general, I will discuss what I believe to be the most important opportunity since the industrial revolution, rivalling it in significance. This unique circumstance is created by the emerging Very Large Scale Integrated (VLSI) technology, with which enormously complex digital electronic systems can be fabricated on a single chip of Silicon one-tenth the size of a postage stamp. Out of it systems will be created which radically change our modes of communication, commerce, education, entertainment, science and the underlying rate of cultural evolution. The quality of human life can be improved in remarkable ways by these changes. Electronics creates no noxious by-products and uses only miniscule amounts of energy. It can accomplish tasks which were previously energy intensive, and dangerous or degrading to human workers. There is no doubt that this electronic revolution will take place.

For 20 years the US semiconductor industry has been the world's prime example of innovative excellence. Each new round of technology was carried by a new wave of start-up companies financed with venture capital. These in turn became the mainstays of their market only to spawn another brood as a new opportunity unfolded. Each turn of this wheel of fortune added new markets, formed new capital, and created additional jobs. The spirit of those heady times is by no means dead but the basic nature of the game has changed. System design, not technology, is now the area in which small firms outshine their giant mentors. Implementation of such designs, however, requires a silicon wafer-fabrication capability unaffordable by any single small enterprise. A fabrication facility serving many such firms can set the industry on a course even more exciting than that of earlier times. In what follows I have tried to describe the status of the field and how a simple, decisive act can assure a dynamic new round of innovation within it.

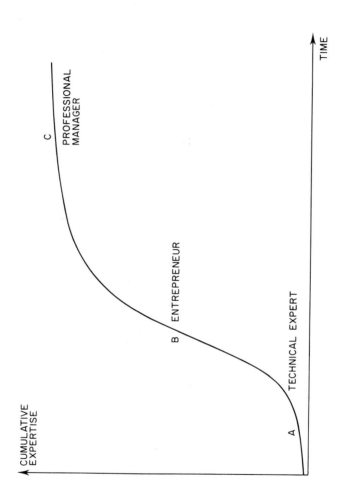

INNOVATION

I have found the following simple view helpful in understanding how innovation has worked in the past, and how we might encourage it in the future. The semiconductor technology is composed of a set of disciplines which must be considered separately. In any given discipline, innovation proceeds along an S-shaped curve such as that shown in Fig. 1. In the early phases, marked (A) in the figure, progress is limited by the lack of fundamental ideas. A single good idea can make possible several other good ideas and hence the innovation rate is exponential. During this period, a single individual or small group of individuals can develop a viewpoint and contribute several crucial insights that set a field in an entirely new direction. It is the time during which progress is dependent upon a few visionaries within the field. During the central and most visible portion of the evolution, marked (B) in Fig. 1, a linear region ensues. Here, the fundamental ideas are in place and innovation concerns itself with filling in the interstices between these ideas. Commercial exploitation abounds during this period. Specific designs, market applications, manufacturing methods grow rapidly. The field has not yet settled down at this point. Entrepreneurs backed by venture capital firms can have a large impact and achieve a dominant market share during this period. During the later stages of the evolution curve, marked (C) in Fig. 1, progress becomes logarithmic in time. Few changes in the market share "pecking order" occur. Manufacturing methods are refined even further. More and more capital is expended to reduce the price of manufacturing. Here, the business becomes capital intensive. Production know-how and financial expertise are required credentials. Professional managers and large firms dominate the business.

THE ARENA

VLSI is a statement about system complexity, not about transistor size or circuit performance. VLSI defines a technology capable of creating systems so complicated that coping with the raw complexity overwhelms all other difficulties. From this definition, we can see that the way in which the industry responds to VLSI must, in fact, be different from the way it has historically evolved through its other phases.

The complexity scale implied by the new technology can be appreciated from the analogy (Setz, 1979) presented in Fig. 2. At several points in the evolution of the technology,

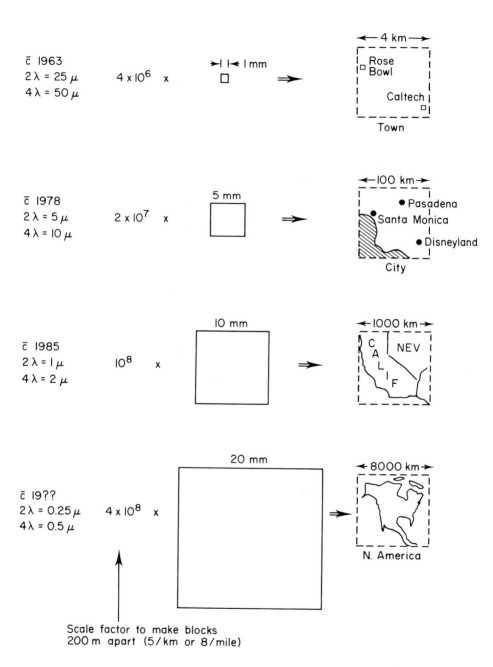

Fig. 2.

a typical chip has been scaled up to make the spacing between conductors equal to one city block. The circuit can then be thought of as a multi-level road network carrying electrical signals instead of cars. In the mid 1960s, the complexity of a chip was comparable to that of the street network of a small town. Most people can navigate such a network by memory without difficulty. Today's microprocessor using a 5µ tech-technology is comparable to the entire Los Angeles basin. By the time a 1µ technology is solidly in place, designing a chip will be comparable to planning a street network covering all of California and Nevada at urban densities. The ultimate $\frac{1}{4}$µ technology will likely be capable of producing chips with the complexity of an urban network covering the entire North American continent. Designers are just now beginning to face complexity as a central and dominant issue of the next stage of evolution. In order to realize the full potential implied by such complexity, entirely new design methods and system organizations must be invented. A high rate of innovation is required to achieve leadership in this remarkable arena.

The evolution of the component fields which make up the present VLSI disciplines are shown schematically in Fig. 3. Progress in each depends upon those before it being well in place. By now, the number of dramatically new ideas being added to the device physics area is small. Fabrication technology has essentially all the fundamental knowledge that will be required. Circuit and logic design have some clever-ness left but that too will soon saturate. The large system design methodology is still in its exponential phase. Many fundamental ideas have yet to be discovered. The organization and programming of highly concurrent systems are even less well developed. Only a few results are known, and much of the fundamental conceptual apparatus needs to be discovered. A period of very rapid growth lies ahead of us in both of these disciplines. They are central to the difference between VLSI and the current way semiconductor devices are designed. It is here that the major innovative possibilities lie.

Historically, innovation in the industry has been spear-headed by small start-up firms and later taken up by large existing organizations. It is significant that the major suppliers of vacuum tubes did not become the major suppliers of transistors. The major suppliers of discrete transistors did not give us semiconductor memories. More recently, companies dominant in the semiconductor memory business did not bring us the multiplexed address random access memory. The microprocessor did not come from mainframe or minicomputer firms. Each of these innovations was brought to market fruit-ion by a small start-up firm which rapidly gained market

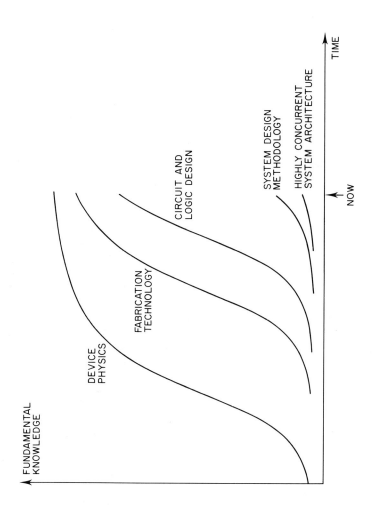

share by virtue of its innovation. Existing dominant firms were then forced to retrofit these ideas these ideas into their own product lines. For the reasons mentioned above, I expect this trend to be even more apparent as VLSI evolves.

The problem faced by the semiconductor industry is thus apparent. Fabrication technology has reached its capital intensive stage. Design is still very early in its exponential phase. Each small system group can no longer afford its own fabrication area. A start-up firm of ten years ago with a capital budget of one or two million dollars for a fabrication area was within the means of traditional venture capital sources. However, the same is not true for capital budgets of several tens of millions of dollars required for state-of-the-art fabrication lines in the near future. If innovation by a myriad of small groups and individuals is to carry us into the VLSI revolution, we must not expect these groups and individuals to provide their own fabrication facilities. The level of innovation required can be achieved only if fabrication is provided as a service by a few well capitalized firms.

Every time a qualitatively new element has been introduced into the industry, new business opportunities have been created. Small firms have obtained significant market shares by extending markets previously dominated by larger firms. The VLSI revolution we are facing is no exception. I fully expect a very large number of small firms to create entirely new machine organizations and entirely new design methodologies. These will allow small, able groups to succeed in the varied market place for systems in spite of historic dominance by capital intensive computer and semiconductor houses.

The seeds of the new wave of innovation have already begun to take root. Much of the thrust is coming, not from the integrated circuit industry itself, but from the collaborative effort of faculty members in many university computer science departments and scattered individuals throughout the industry.

A new type of course is being offered at these universities that provides students of computer science and electrical engineering with a thorough introduction to integrated system architecture and design. The courses provide the minimum of basic information about devices, circuits, fabrication technology, logic design techniques, and system architecture, which is sufficient to enable the student to span fully the entire range of abstractions, from the underlying physics to complete VLSI digital computer systems. Only a surprisingly small set of carefully selected key concepts is necessary for this purpose. By minimizing the mental baggage carried along each step, the student emerges with a good overall understanding of the subject. The courses are based on a course I originated at Caltech in 1971, and on a new and highly integrated system design methodology that Lynn Conway (of the Xerox Palo Alto Research Center) and I present in a major new textbook on the subject (Mead and Conway, 1978/79).

A major portion of these courses is devoted to "learning by doing", with the student undertaking the architecture design, layout and testing of substantial LSI system project A typical student project might contain on the order of several hundred to several thousand transistors. The students typically describe their designs using a symbolic layout language. Only modest facilities are needed to support such courses, in addition to that commonly available at these universities. The universities do not usually have their own LSI fabrication facilities. The "multiproject chips" containing all the LSI projects for such a course are instead fabricated by commercial maskmaking and wafer fabrication firms, directly from the design produced by the students.

Throughout the evolution of courses and research projects, we have constantly been hampered by the difficulty of gaining access to wafer fabrication service. Although many industrial firms have been very helpful in this regard, their facilities are dedicated to their own proprietary products, and were made available to us on a special case, special favor basis. While these favors were crucial to getting the university programs going, they obviously cannot be the basis for supplying a multitude of emerging small system firms, nor can they effectively support the expanding group of participating universities. What is needed is an openly available service facility which will, for a fee, accept pattern files in a standard format (which we have already worked out) and "print" them on silicon wafers using a standard fabrication process.

The analogy with printing is a particularly apt one (Conway While wafer fabrication is the most sophisticated manufacturing technology ever undertaken by the human race, it is pattern independent. A limitless number of system designs can be replicated by a single process, much as books on any number of subjects can be replicated by a single printing press. Individual designers require access to fabrication in exactly the same way that individual authors require access to printing.

An open fabrication service should be established as the collaboration of an industrial partner with the university community. Funding arrangements should be left flexible, and a rigid format should not be prescribed. Once one or two of such facilities have been operating for a year or two, demand will increase and they should become profitable business enterprises in their own right. It is important that the only restriction placed on these facilities is that they make fabrication service openly and uniformly available. It is perhaps the only way to assure Freedom of the "Silicon Press" (Conway).

REFERENCES

Seitz, Charles (1979). "Self-Timed VLSI Systems", from the Proceedings of Caltech Conference on VLSI, January 1979.
Mead, Carver and Conway, Lynn (1978/79). "Introduction to VLSI Systems". Limited printing by the authors, 1978; Addison-Wesley, 1979.
Conway, Lynn. Personal communication.

GENERALIZED IC LAYOUT RULES AND LAYOUT REPRESENTATIONS

Carlo H. Séquin

Computer Science Division
Electrical Engineering and Computer Sciences
University of California, Berkeley, CA 94720

1. LAMBDA - BASED LAYOUT RULES

The evolution of very large scale integration (VLSI) is accompanied by a proliferation of ever more sophisticated and thus often more complicated processes. At the same time the cost of the design and layout of a VLSI circuit has become so prohibitive that one can simply not afford to hand-code the layout of a particular circuit for every process variant. A simpler and portable set of layout representations is thus highly desirable. Lyon (1981) has recently given an extensive introduction and description of a set of simplified design rules for the prevailing silicon-gate NMOS process. These simple rules, introduced by Mead and Conway (1980) and extensively used in the context of the multi-project-chip efforts (Conway *et al.* 1980), are based on a single parameter *Lambda* (λ) which makes the rules scalable and thus potentially gives them a much longer lifetime. As long as technology makes nearly uniform advances on all fronts, i.e. worst case misalignment and the smallest feasible features in all mask levels scale down at roughly the same pace, these rules and the associated layouts maintain their validity for all but the most stringent demands.

While longevity is clearly an important attribute of these rules, there are other advantages too: These rules are very simple and thus much more suitable for novice designers. They can be readily expressed on a single color plate, whereas a typical set of industrial design rules comprises more than a hundred separate rules and easily spans a dozen pages of instructions and sketches. But even for the experienced designer there are advantages. By using these simple layout rules and placing all features on a fixed *Lambda grid*, one can work more effectively, uncluttered by the nitty-gritty details of the ultimate fabrication rules, and focus on the more important higher level aspects of the layout. Even if the layout is partially done by a computer, the simpler rules have the advantage that they can reduce the computational task in circuit verification and layout rule checking.

2. EXTENSION TO OTHER PROCESSES

This paper presents a simple set of generalized rules that are formalized independently of a particular fabrication sequence, emphasizing those tolerances that are common to many different processes. The set is simple, regular, and easy to remember. It can be used as a starting point for the more sophisticated set of layout rules of a specific fabrication process by adding a - hopefully rather short - list of well justified exceptions.

The simplicity of the generalized set of rules has advantages for the designer as well as for the developer of efficient design tools. This approach using fixed rules with added exception lists has the further didactic advantage that it explicitly points to the differences between various processes and thus highlights potential trouble spots.

All dimensional rules are based again on a single parameter *Lambda* (λ) that characterizes the linear feature "resolution" of the complete wafer implementation process and permits first order scaling. These layout rules are subdivided into three groups of less than ten rules each, concerning respectively: Mask Feature Sizes, Overlaps and Separations, and Macroscopic Rules associated with the chip periphery and bonding pads.

3. MEANING OF A LAYOUT - CONCEPTUAL LAYOUT LEVELS

The more sophisticated CMOS and bipolar processes are often so complicated that it becomes inefficient for the designer to think in terms of *all* the mask levels actually used in the fabrication process, many of which may only be artifacts of a particular implementation sequence. The design process can be abstracted to a few conceptual layout levels that represent the relevant physical features to be found in the final silicon wafer. The actual mask levels required for implementation, which may vary for different fabrication lines, can then be computer generated from these conceptual design levels.

For example, rather than dealing with one or two thin-oxide masks, channel stop masks, n- and p-type implant masks, as may be used in a typical silicon-gate bulk CMOS process, the designer should think in terms of the $n+$ and $p+$ regions that form the operational source and drain electrodes of the transistors of both polarities. Thus the designer may draw only a subset of the masks required for actual manufacturing, or perhaps a composite of two or more layers. An example is the "green" areas in the silicon-gate NMOS process used by Mead and Conway (1980). While these areas are referred to as "diffusion", they really describe the thin oxide areas; the actual diffusion or implant, which is applied uniformly across the whole wafer, is being properly masked by the thick oxide and by the polysilicon features present at that time.

The introduction of such conceptual levels raises the important question of what the layout actually represents:
- a) Is it a scaled picture of the mask geometries ?
- b) Does it correspond to the final features on the wafer ?
- c) Is it an idealized geometrical representation of the intent of the designer in terms of certain device parameters ?

We advocate that, for the kind of design environment and methodology that such a set of generalized layout rules reflects, the most appropriate interpretation lies somewhere between options b) and c). While it is desirable to have the designer think in terms of the envisioned results rather than the means of achieving them, a direct specification of a layout in terms of the final device parameters is beyond the state of the art of current implementation technology. Interpretation c) would understand the "channel width" of a MOS transistor as a scalable parameter which is strictly proportional to the drain current of the device. Since such an ideal parameterization is possible only in a few simple cases, this approach is not practical in general. Nevertheless, the designers should be shielded from the detailed steps of the implementation process and should be able to think in terms of the features they will find on the finished chips that are returned to them. The computer can be used to generate the necessary transformations and combinations of these desired geometrical device features in order to derive the required mask geometries for their fabrication.

3.1. A Comprehensive Set of Levels

At a high enough conceptual level all MOS processes use some of these same basic features: A maximum of two different substrate areas of opposite polarity; strongly doped areas of both polarities forming the source/drain regions of both types of transistors and the contacts or guard rings to the substrate or to the wells; some implants to adjust the threshold of groups of select transistors; gate electrodes; interconnection runs; and contact windows between different conducting levels. Thus a relatively small set of conceptual layout levels depicting these basic features should be sufficient to outline the geometry of any MOS circuit. The following set of conceptual mask levels is believed to be sufficient for all silicon-gate MOS and CMOS processes.

CIF name	representing
ANWL, APWL	Well area in the bulk or type of island doping in CMOS SOS
ACAP	Heavily doped area for capacitor electrodes
AND	N-doped area for n-channel source/drains
APD	P-doped area for p-channel source/drains
AIIN, AIIP	N or P-type implant to adjust FET thresholds
AII2, AII3	even more implant levels ...
ASI	Poly-Si electrodes
AME	Metal interconnects and pads
ACC	Contact cut
ABC	Buried contact
AOC	Overglass cut

To illustrate the difference between these abstract levels and the actual fabrication mask geometries, we will discuss the simple case of ANWL, the outline of the final n-type well in the p-type substrate. We assume that during wafer processing the dopants forming this well diffuse laterally to a distance of 2λ. Thus, while a minimum diffusion mask window of 2λ can readily be implemented, the minimum feasible width of a strip of n-well diffusion will be

6 λ; and if the diffusion mask windows are separated by less than 6 λ, neighboring strips may accidentally merge. The designer working with the conceptual levels will thus be told that minimum feature dimension for this particular layer is 6 λ, and that neighboring strips must be separated by at least 2 λ to guarantee geometrical separation. The computer will then shrink by 2 λ, the amount of the lateral outdiffusion, the wide features laid out by the designer.

3.2. Usage of Conceptual Levels

These conceptual CIF levels are an idealized representation of the salient features to be found in the final silicon wafer. Some levels are specifically related to a particular process class: bulk processes have to deal with the well of opposite polarity containing one type of transistors; metal gate processes must use a separate mask to define the thin-oxide areas forming the active gate regions. Obviously a single geometrical layout cannot be used for all possible CMOS processes, but it may be sufficient for most processes in a particular class such as Si-gate bulk CMOS or CMOS-SOS. Every mask required in a particular processing step is then derived from the set of conceptual levels. It might even be possible to use the same layout for both classes of silicon-gate bulk CMOS processes; the role of well and substrate would simply be reversed. Of course the electrical characteristics of the transistors generated by the two complementary processes may be quite different, and such portability can thus be achieved only for uncritical digital circuits.

The use of these abstract levels also has advantages for the design tools; working with fewer levels will increase their efficiency. The simple description in terms of a few conceptual levels is a suitable intermediate form between a layout done at the sticks level and the final, compacted mask geometry. A two-step compaction from sticks to final layout may be more efficient. In the first steps that yield the dominant amount of compaction, one should consider only the conceptual levels since this means less computational work for the processor. The geometrical layout at the conceptual level also permits verification and simulation of the circuit. Transistor ratios, the resistance of interconnections, and all other salient circuit parameters can readily be extracted from the features in these design levels. These simple layouts can either be used to generate relatively relaxed implementations of a particular circuit by extending them "in place" to the full mask set, or they can be used in conjunction with a circuit compaction program to generate a dense layout with a sophisticated set of design rules.

4. META RULES AND GENERALIZED LAYOUT RULES

At the highest conceptual level, a single design rule can be formalized from which all specific design rules can be derived:

> Under worst case misalignment and maximum edge shift of any feature no serious performance loss should occur.

However this rule is too general to be of much help, and in order to derive more specific rules, a lot must be known about the processing details. So, two assumptions will be made, to permit a derivation of rules of a more specific nature.

1. Assume: worst case misalignment of any two levels is λ.
2. Assume: maximum edge shift due to processing is $\frac{1}{2} \lambda$.

With these two assumptions, a set of rather general, parameterized rules can be derived. They have been grouped together by the nature of one or two related issues that they address.

4.1. Mask Feature Sizes

This section addresses dimensional rules that are concerned primarily with a single mask or feature level and specify minimum internal and external separation of edges. Everything follows form one simple metarule:

MI. *All dimensions must be at least 2λ.*

This rule can readily be expanded into the following more specific rules which are also illustrated in Figure 1:

1. Minimum feature size in any level is 2λ.
2. Minimum separation between features is 2λ.
3. Minimum contact window size is $2 \lambda \times 2 \lambda$.
4. Minimum contact cut separation is 2λ.
5. Minimum conductor width is 2λ.

Rules 1 through 4 follow directly from metarule MI. Rule 5 also follows in the same way if it is understood to refer to a conductor such as a polysilicon path defined by a single mask level; however it can be generalized to mean that *any* conducting path should be at least 2λ wide, even if it is defined jointly by two mask levels, as is the case in a diffused path running along a polysilicon gate.

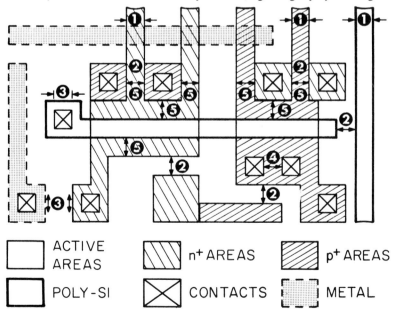

Fig. 1. Rules on mask feature sizes.

4.2. Overlaps and Separations

The underlying meta-rule that governs the overlaps and separation of features on two different masks is:

MII. *Features that should not touch must be separated:*
- *by two λ - if touching has a catastrophic effect,*
- *by one λ - if touching has NO catastrophic effect.*

From this the following rules can be derived (see Figure 2):
1. Frames around contact cuts must be at least 1 λ wide.
2. Keep contact cuts above an active layer 2 λ from edge.
3. Keep contact cuts 2 λ away from poly-Si gates.
4. Separate poly-Si and diffused conductors by 1 λ.
5. FET-gates must extend 2 λ beyond the transistor channel.
6. Doping of a given area should be extended by 1.5 λ.
7. Keep doping 1.5 λ away from undesired areas.
8. Buried contact window frames must be 1.5 λ.
9. Keep buried contacts 1.5 λ away from undesired areas.

Rules 8 and 9 are the simplest way to deal with buried contacts. They correspond in spirit to rules 6 and 7 dealing with threshold adjusting implants. A more detailed analysis is given by Lyon (1981). However, because of the large variety of possible layout configurations of buried contacts, a complete catalog of all cases is beyond the scope of this list.

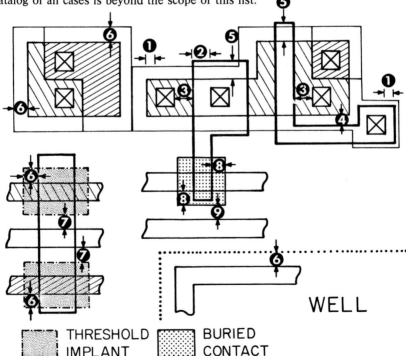

Fig. 2. Rules on overlaps and separations.

4.3. Macroscopic Features

Some features on an IC chip are not normally scaled down proportionally to λ. Features such as the scribe lines between individual chips on the wafer, or size and separation of bonding pads, are really determined by the dimension of the mechanical tools used in dicing and interconnecting these chips. Since the technological progress in this area is slow compared to the advances in photolithography and chip fabrication, the associated rules are expressed in absolute quantities:

1. Total scribe line width is 100 μm.
2. Minimum feature distance from scribe line is 50 μm.
3. Minimum bonding pad size is 120 μm.
4. Minimum overglass cut to bonding pad is 100 μm.
5. Minimum feature distance from bonding pad is 40 μm.
6. Minimum bonding pad separation is 80 μm.
7. Minimum bonding pad pitch is 200 μm.

Rules 3 and 6 result in a bonding pad pitch of 200 μm. This is desirable so that the chips can be wire-bonded easily and with good yield. A few additional constraints sound rather obvious but are often overlooked by the novice designer:

8. The final chip must fit into the cavity of the package.
9. The length of any bonding wire should not exceed 5 mm.
10. Bonding wires must not cross.

Rules 9 and 10 imply that the pads on the chip should roughly line up with the pads in the package and should be spread evenly around the chip perimeter. Thus for use in a 40-pin package one should not put more than a dozen pads along any one edge of the chip.

4.4. Derivation of NMOS Rules

To illustrate the concept of *exceptions* that produce a usable set of rules for a real process, we present the necessary additional rules for the Si-gate NMOS process referred to by Mead and Conway (1980). Only two rules need to be added to obtain their set of layout rules:

a. Transitions from thin gate-oxide to thick field-oxide in the local oxidation processes use up space. The minimum separation of thin-oxide areas thus is 3 λ. In other words, minimum feature size of thick oxide is 3 λ.

b. Metal lines on non-uniform surfaces have poorer edge definition and should therefore have coarser features; a minimum width and minimum separation of 3 λ are often specified.

In addition there are some electrical design rules that apply to most processes, such as:

1. Current density in metal must be less than 10^5A/cm^2.

Space constraints do not permit us to discuss them here in more detail.

5. APPLICATION TO CMOS

The above rules and unified layout representations have been generated after review of two Si-gate CMOS processes, a metal-gate CMOS process, two NMOS processes, and a CMOS SOS process. Compatibility with the basic silicon gate NMOS process used by Mead and Conway (1980) and with the CMOS SOS design rules proposed in the Caltech/JPL standards project (Griswold 1980) has been maintained as much as possible.

The design approach using the above proposed conceptual levels will now be illustrated for the inverted bulk CMOS process, chosen for its generality and significance for the future of VLSI. It is a direct and compatible enhancement of the Si-gate NMOS process that permits to mix high-density NMOS circuits and low-power CMOS circuits on the same chip. The process holds the promise of being scalable to channel lengths of only 0.5 μ. The fabrication sequence outlined here follows a process recently developed at Berkeley (Choi 1981). This process can also provide good MOS capacitors by introducing special n+ regions that act as bottom plates prior to the deposition of the polysilicon top plates.

5.1. Process Outline for Inverted Bulk CMOS

The first mask, MNWL, defines the phosphorus doped n-well. An optional mask, MNCA, then defines n+ capacitor plates in the p-type substrate. The gate oxide is grown, and the wafer is covered with silicon nitride. Mask MAAN delineates the thin-oxide, active areas above the p-substrate but leaves all n-well areas covered. The nitride layer is selectively removed where thick field oxide is desired, and a boron implant is introduced to act as a self-aligned p-type channel stop. After a first field oxidation, a separate mask MAAP then removes more of the same nitride layer, this time above the n-well. Photoresist protects the previously defined nitride islands on the p-substrate during the phosphorus implant that forms the n-type channel stop in the n-well. A long local oxidation step produces the thick field oxides above the p-substrate and the n-well. All nitride can now be removed.

After a threshold adjusting implant, heavily n-doped polysilicon is deposited over the whole wafer. Mask MSIN is used to form the gate electrodes of the n-channel transistors. An arsenic implant will then form n+ regions in all thin-oxide regions no longer covered by poly-Si. During this step the future PMOS devices are shielded by polysilicon. Similarly, a second masking step (MSIP) on the same polysilicon layer defines holes for the p+ regions, i.e. the source and drain regions for the p-type transistors in the n-well and the contact points to the substrate. The photoresist layer of this masking step shields the n+ regions during the Boron implant.

After implant drive-in and annealing, a passivation layer of polyimide is applied. Contact holes defined by mask MCC are cut through polyimide and oxide. Aluminum is deposited and patterned by mask MME. If a protective layer is employed, mask MOC is required to provide access to the bonding pads.

5.2. Mask Level Summary for Inverted Bulk CMOS

The masks that need to be generated for this process and their relationship to the CIF levels specified by the designer are:

Fabrication		Design
MNWL	Well definition	ANWL, shrunk for outdiffusion
MPCA	Capacitor areas	ACAP
MAAN	NMOS active area	AND ∪ APD ∪ ANWL
MAAP	PMOS active area	APD ∪ AND ∪ ¬ANWL
MSIN	NMOS poly-Si gates	ASI ∪ (ANWL ∩ ¬AND)
MSIP	PMOS poly-Si gates	ASI ∪ (¬ANWL ∩ ¬APD)
MCC	Contact windows	ACC
MME	Metallization	AME
MOC	Overglass windows	AOC

The major differences between the set of design and fabrication levels is that the designer deals only with one layer of polysilicon (ASI); the separate masks MSIN and MSIP required for fabrication can be derived from logical combinations of the well geometry and the n+ or p+ areas. This results in a simpler and more meaningful layout.

5.3. Additional Design Rules for Inverted Bulk CMOS

Starting from the generalized design rules outlined in section 4 of this paper, the following additional rules are necessary to obtain a complete set for the above process.

1. As in all processes using local oxidation, the transitions from thick to thin oxide uses up some space; these areas should thus be separated by 3 λ. At the conceptual level this refers to the placement of separate n+ or p+ areas.

2. Well depth and lateral diffusion can vary anywhere from 1 to 3 λ in different processes. Because of this variation, tolerances for this feature must be rather relaxed. Minimum distances to the edge of this layer are typically specified as 4 λ. Thus the following additional rules result:
   - -a. Minimum feature size for level ANWL is 6 λ.
   - -b. Separation between unrelated wells is 4 λ.
   - -c. Distance of unrelated doped areas from well edge is 4 λ.
   - -d. Guard ring is 4 λ wide; 1 λ outside, 3 λ inside ANWL.

3. The PMOS poly gate should be 3 λ, because of the larger boron outdiffusion.

4. Rules concerning the alignment of the shield masks used in the local oxidation steps of the well and substrate areas can be ignored by the designers who work with the conceptual levels; they are contained implicitly in the rules that govern the minimum distance of features from the well boundaries.

5   There are electrical rules, designed to prevent latch-up of the CMOS circuitry. Latch-up is typically a problem only in the peripheral circuits, which should be designed by an experienced designer and placed into a cell library. The user of these cells need not know these rules.

Note in particular, that in this process, because of the smooth surface of the passivation layer, metal lines need not be wider than the minimum width specified by the generalized design rules, i.e. 2 λ.

6. LAYOUT REPRESENTATION

With the widespread acceptance of CIF2.0 (Sproull and Lyon 1980, Hon and Séquin 1980) as a de facto standard for the description of IC layouts, it is important to standardize not only the syntax of this interchange format but also its semantics.

Known mask names		*features*
used in fabrication	*used in design*	*appearing in Si-gate MOS processes*
"MPWL"	"APWL"	P-Well mask / actual extension
"MNWL"	"ANWL"	N-Well mask / actual extension
"MNCA"	"ANCA"	N-type Capacitor Area
"MPCA"	"APCA"	P-type Capacitor Area
---	"ACAP"	Capacitor of either polarity
"MAA"	---	Active Area for any device
"MAAG"	"AAAG"	Active Area under metal gates
"MAAN", "ND"	---	Active Area for N-channel devices
"MAAP"	---	Active Area for P-channel devices
"MND"	---	N-type Doping
"MPD"	---	P-type Doping
---	"AND"	N-type Doped active areas
---	"APD"	P-type Doped active areas
"MIIN", "NI"	"AIIN"	Ion-Implant, N-type
"MIIP"	"AIIP"	Ion-Implant, P-type
"MSI",	"ASI"	Si-gate for any device
"MSIN", "NP"	---	SI-gate for N-channel devices
"MSIP"	---	SI-gate for P-channel devices
"MSI2"	"ASI2"	second level poly-SI for the future
"MSI3"	"ASI3"	third level poly-SI for the future
"MBC", "NB"	"ABC"	Buried Contact
"MCC", "NC"	"ACC"	Contact Cut
"MCC2"	---	higher level or oversize Cut
"MOC", "NG"	"AOC"	Overglass Cut mask
"MME", "NM"	"AME"	MEtal mask
"MME2"	"AME2"	MEtal mask, second level
"MME3"	"AME3"	MEtal mask, third level

6.1. Mask Level Names

In addition to the already established seven mask layer names for NMOS, we propose here an additional two set of names. One set, in which all layer names start with "M", is a comprehensive and expandable set of mask levels used for the actual device fabrication. The other set, containing names starting with "A", denotes the abstract design levels discussed earlier in this paper.

6.2. Layout Colors and Stipple Patterns

The introduction of some standard colors by Mead and Conway (1980) and the acceptance of these colors by a large University community has simplified dramatically the communication between the designers adhering to this particular representation. It is desirable to develop such a shared culture for a much wider set of processes. Unfortunately, already entrenched conventions and the different limitations of various output devices (screen or plotter) may render this a futile dream. Nevertheless we have worked out a usable set of colors spanning all classes of MOS processes, and we have also done work in developing a readable set of black and white stipple patterns. Restrictions on space do not permit us to present this work here. Interested readers should ask for a copy of the extended version of this paper.

ACKNOWLEDGMENTS

This work was sponsored in part by the Defense Advance Research Projects Agency and performed in part under a consulting contract with Xerox Palo Alto Research Center. I would like to acknowledge many fruitful discussions with R.F. Lyon.

REFERENCES

Choi, T. (1981). "The inverted CMOS process ...," *in preparation.*

Conway, L.A., Bell, A., and Newell, M.E. (1980). "MPC79: A Large-Scale Demonstration of a New Way to Create Systems in Silicon," *Lambda,* Vol 1, No 2, pp 10-19.

Griswold, T.W. (1980). "CMOS SOS Design Rules," CALTECH/JPL Standards Project.

Hon, B. and Séquin, C.H. (1980). *A Guide to LSI Implementation,* Second Edition, Xerox PARC, Palo Alto, CA, Jan 1980.

Lyon, R.F. (1981). "Simplified Design Rules for VLSI Layouts," *Lambda,* Vol 2, No 1, pp 54-59.

Mead, C.A. and Conway, L.A. (1980). *Introduction to VLSI Systems,* Addison Wesley, 1980.

Sproull, R.F. and Lyon, R.F. (1980). "The Caltech Intermediate Form for LSI Layout Description," in: *Introduction to VLSI Systems,* by Mead, C.A. and Conway, L.A., Addison Wesley, pp 115-127.

A NON-METRIC DESIGN METHODOLOGY FOR VLSI

E. E. Barton

*Inmos Ltd, Whitefriars, Lewins Mead,
Bristol BS1 2NP*

1. INTRODUCTION

The ability to fabricate integrated circuits of great complexity has increased the difficulties of designing large scale integrated systems. Research into the design of such systems has attempted to lighten this burden by the introduction of various design aids. These aids have however, limited themselves mostly to the spatial layout of a circuit on the surface of a piece of silicon. There exists another source of complexity in the design of large systems, namely time.
Initially, the layout of integrated circuits was firmly linked to a spatial metric, by the specification of the layout geometry in terms of absolute coordinates on a fixed grid. Scaleable design rules (Mead and Conway 1979) loosened this bond and aided the implementation independent design of systems, however the spatial metric still formed the foundation on which designs were based. Two functionally equivalent cells require different layouts depending on the context in which they were placed. Stick diagrams (Williams 1977) reduce the need for such a rigid design methodology, by retaining only the information which is essential to the correct implementation of circuits, namely an abstraction of the geometry of the components and a partial ordering of their relative positions in two dimensions. A cell design therefore becomes independent both of the fabrication techniques used in its implementation, and the local context in which it is placed in the total design.
The use of hierarchies in the design of integrated systems helps to partition a design into manageable "chunks" which a single designer can encompass, thereby encouraging locality

and modularity. The further partitioning of the hierarchy
into "leaf cells" and "composition cells" (Rowson 1980) aids
its analysis and reduces all metric-based representations to
the lowest level cells.

The design of synchronous systems may be likened to a
metric-based layout methodology. In that signal events may
occur only at certain pre-determined discrete intervals in
time. The step to self timed systems removes this stipula-
tion in much the same way as stick diagrams facilitate the
design of systems in space, by specifying only the sequence
in which events may occur, and not the length of the intervals
between them. In addition, the design of self-timed systems
fits a hierarchy in which only the lowest levels need to be
concerned with the implementation of the sequence specification in terms of timing constraints. Above this level,
larger systems are built from legal interconnections of self-
timed systems under a set of composition rules which
guarantee correctness, thereby reducing the complexity of
verification.

2. PRINCIPLES

The distribution and organisation of clocking signals has
not been a major problem in current designs. If chip sizes
remain roughly constant, but feature sizes scale down, the
lengths of the wires used to implement global clocking
schemes will become a lot longer in proportion to the sizes
of gates. As the propagation delay increases quad-
ratically with reduced feature sizes, diffusion delays will
become appreciable and the delays incurred when driving
global signals round a large chip will require very careful
consideration.

In a self timed signalling convention, pairs of signals
are partially ordered by the precedes relation ($<=$). If
the occurence of two signals, A and B are related by this
operator ($A <= B$) then A precedes B. The concept of sim-
ultaneity is ruled out by the anti-symmetric property of
$<=$. Any two signal events which are not related by $<=$ are
said to be concurrent or unordered.

The determination of the sequence of signal events is made
locally by the lowest level circuits of a self timed system,
named self timed elements. It is only in the self timed
elements, that the sequence domain is embedded in time. The
problem of diffusion delays in wires may be minimised by
approximating regions over which their effects may be ignored, termed "equi-potential regions". Self timed elements
are constrained to lie completely within at least one equi-
potential region.

Larger self timed systems are made up of a number of smaller self timed systems, interconnected under a set of composition rules. These composition rules are purely concerned with the correct sequencing of signal events between the smaller self timed systems, as the mapping from sequence to time is already accomplished by the self timed elements. Computation therefore proceeds by the cooperation of communicating sets of self timed systems, rather than the imperative commands of a set of clocking signals.

The rest of this paper is about an attempt to apply the principles of self timed design to the implementation of a real system.

3. IMPLEMENTING A SELF TIMED SYSTEM

Initially, the design of a PDP8 CPU was attempted, to prove that larger systems could be designed with self timed logic. An early decision was to use a delay insensitive signalling scheme for two reasons. Firstly, this signalling scheme puts the self timed elements, down at the level of individual bits and gives a worst case measure of the area required by self timed designs. Secondly, all subsets of the design could, for initial testing, be implemented on separate chips and delays in driving signals off and on chip would be no problem.

Signalling Scheme

A delay insensitive signalling scheme (Mead and Conway 1979) was used for the CPU, implemented as a negative double-rail code. This form was used because pass transistor networks are easily used to implement switching networks, with pull-ups on the outputs of both wires to keep the signal undefined, until a result is computed. Then one wire will be pulled low, and the output will become defined. An additional reason for using this code was the ratioed nature of Nmos logic. A self timed element works in two phases, the first of which computes a result from some inputs, and the second of which must propagate undefined signals to all parts of the element. Whereas the first phase may involve some longer delays, such as passing on carry signals, the second can usually proceed faster. It is therefore sensible to associate downward going transitions on wires during the first phase, and upward going transitions during the second.

Building Blocks

The first stage of the design was to implement the ALU. A first attempt at microcoding the instruction set, revealed that the ALU would best be implemented as a pipeline in order that register transfers could proceed concurrently with ALU operations. The ALU therefore consists of two pipeline modules and a block of combinational logic (CL). The first pipeline module latches incoming data from two busses along with an OP code and the second latches the result of the ALU operation. The CL implements addition, rotation, ANDing and ORing.

Pipeline module

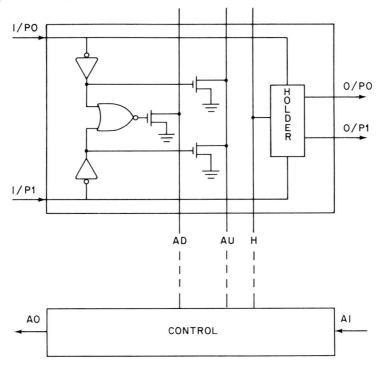

Figure 1

The design of the pipeline module illustrates many of the techniques used to implement self timed systems and is therefore described in some detail. The pipeline module (Fig. 1) consists of a number of cells which latch data signals ("Holder"), and a control state machine ("Control"). Three wires run the length of the pipeline module. "All defined"

(AD) and "All undefined" (AU) are used to detect the state of the inputs. "Hold" (H) directs the Holder cells to latch the data. The control cell has two additional signals, "Acknowledge in" (AI) and "Acknowledge out" (AO). The first of these is used by the environment to signal the reception of data from the module, and the second is used by the module to signal data acceptance to the environment.

Before a self timed element can be implemented, there must exist a rigid specification of the sequence of the signal events occuring at its terminals. While this sequence may be specified by the precedes relation, a subset of Petri Nets termed S nets (Seitz 1970), provides a useful graphical alternative. When modelling a circuit, each transition of the S net must be labelled with a single signal event. A marking of the S net corresponds to a certain state of the circuit, thus as tokens move round the net, the firing of the transitions model the circuit's behaviour.

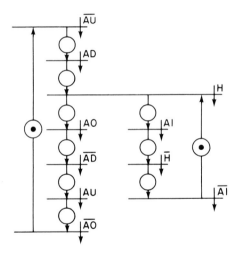

Figure 2

Figure 2 shows the S net of the control cell. Data is initially all undefined, and H is low. As data becomes defined AU drops, until all the input data is defined and AD rises. At this point Control responds by raising H and AO. When the environment observes AO, it will begin to remove defined data and AD will drop. When the data is all undefined, AU goes high and AO is lowered. Concurrently, the environment will have observed defined data on the outputs of the holder cells. When all the outputs are defined, it will raise AI, which in turn causes Control to lower H. This signals the Holder cells to remove the defined output data, when the environment will acknowledge by lowering AI. This has taken Control through one cycle.

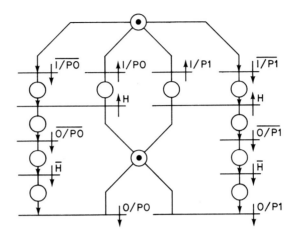

Figure 3

Figure 3 shows the S net of the holder cell. The input and the output signals are initially undefined, with H low. When either of the input wires drops low, H may be raised, and the corresponding output wire goes low. Thereafter, the input wire which dropped may rise again making the input undefined. Concurrently, H may be lowered, indicating that the output should be made undefined. When it has, the next set of data may then be latched.

The reachability set of an S net, found by the vector addition method (Karp and Miller 1969), enumerates all the possible states in which the circuit may find itself, and may be used to detect incorrect S nets. This net is a tree in which the root is the initial marking, the branches represent signal events, and the leaves are either "dead" states, or repeated states found on the path from the root to that leaf. When analysing S nets, the vector addition method may be modified so that as each next state is found, it is checked against the current extent of the tree in order to detect a repetition. If one is found, an arc is inserted from the present state to that state, and the algorithm does not need to search further. When the arcs are substituted by transitions, the result is an S net which models the original with the restriction that there is only one token in the whole net at any time. This S net may then be converted to a conventional state diagram, usually by assigning states to groups of places directly connected by output transitions. This state diagram is then reduced by merging states, under the usual rules applied to asynchronous state machines.

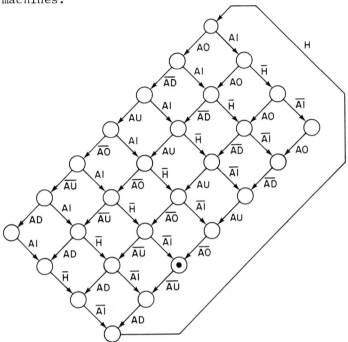

Figure 4

Applying the modified vector addition scheme to the Control S net yields figure 4. This diagram is then converted to a state diagram, state assignments are made and the machine is minimised (Fig. 5). Finally the state equations are produced so that no hazards or race conditions may arise.

The laying out of control may now proceed. An important point to note is that the raising of AO may only occur sufficiently long after the raising of H to ensure that the holder cells actually have latched the incoming data. In the current implementation, the control state machine senses the H line, and can only transit to the next state when this line is at a "good" logic level. The delays in the holder and control circuits have been tuned such that AO will only rise after the holder circuits will have observed H.

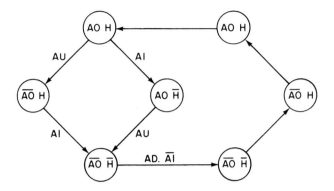

Figure 5

The CL may be divided into three main sections. The first of these is a decoder which takes a three bit OP code and produces one ternary signal which selects between addition and the logical operations, one ternary signal which specified AND or OR, and one six-valued signal which specifies the rotation. The second unit is the adder and logical operations unit and the third is the rotator.

Decoder

The decoder is implemented as a PLA, however some modifications are required to normal PLAs for use in self timed systems. The AND plane of a PLA has an even number of input columns; one of which must be low when the input signal is high, and the other which must be low when the input signal is low. In single rail logic, one of the columns must therefore be driven from its input via an inverter. In negative double-rail logic, the two wires of a signal already encode the required information, therefore for each input wire, there corresponds one input column. The ternary outputs are decoded into two wires each, while the signal controlling the rotator requires five wires.

Adder

Addition is the only operation required of the ALU which involves global communication in implementing the carry path. The ternary signalling scheme has been used to advantage in this implementation to minimise the effective length over which carry signals must be propagated. If a completely random set of operands are assumed, the expected number of consecutive bits which are not equal is the log of the length of the word. If two bits are equal, the carry out for that bit can be calculated immediately (carry kill and carry generate conditions). Several carry signals may therefore propagate concurrently through the adder in most cases and the total time to add the operand is correspondingly reduced.

The first stage of the adder obtains the positive logic signals, carry pass, carry generate and carry kill. The second stage computes the carry out, the third stage computes the sum as the parity of the input signals and the carry in. The sum is identical to the carry when the input bits are equal, and its inverse when they differ. Inverting a double-rail encoded signal is done by swapping the two wires round, therefore the sum may be computed by passing the carry in signal directly to the output when carry kill or carry generate are high, and by swapping the wires round when carry propagate is high.

Logical Operations and Rotation

In order that inputs do not need to traverse the whole adder, AND and OR are computed from the carry generate, kill and propagate signals. This is directly implemented with inverting logic. The rotator is basically a reduced form of crossbar switch, implemented as a pass transistor network.

Silicon Implementation

The ALU was implemented in Nmos with a 5 micron minimum gate width. The total delays in the circuit were calculated to be 94 gate delays for logical operations, 241 in the worst case of addition, where carry must be propagated along the whole length of the adder, and 106 gate delays in the expected case, where the longest path before a kill or generate condition exists, is through three bits. The ALU take 61 gate delays to propagate undefined data throughout its length. The dimensions of the ALU were 50 x 105 mils.

4. CONCLUSIONS

Self timed design is a method of abstraction, to limit the numbers of details on hand during the stages of a design. The design of the PDP8 ALU was done to illustrate the use of this method and to test it out. The method was therefore used right down to the level of individual bits, to test the trade-offs at all levels of the design. The exercise has shown that simplifications in the design process are possible. There are also speed advantages as a circuit's completion signal can initiate the next computation, rather than having to wait for some worst case delay. There is an area penalty, as the self timed elements have to be "smarter" than their synchronous equivalents however this penalty may be minimised by an appropriate choice of the level at which the self timed elements will be implemented.

References

Karp and Miller (1969) "Parallel Program Schemata" *Journal of Computer and Systems Science, 3.*
Mead and Conway (1979). "Introduction to VLSI Systems". Addison-Wesley Publishing Company, Reading Massachusetts.
Rowson (1980). "Understanding Hierarchical Design". PhD Thesis, Dep. of Computer Science, Caltech.
Seitz (1970). "Asynchronous Machines Exhibition Concurrency" Project MAC Conference on Concurrent Systems and Parallel Computation, ACM Conference Record.
Williams (1977). "Sticks-A New Approach to LSI Design", MSEE thesis, Dept, of Electrical Engineering and Computer Science, MIT.

Acknowledgement

This work was carried out while the author was a graduate student at the California Institute of Technology, under the guidance of Charles L Seitz.

TOP DOWN DESIGN OF A ONE CHIP 32-BIT CPU

R. H. Krambeck, D. E. Blahut, H. F. S. Law,
B. W. Colbry, H. C. So, and M. Harrison
J. Soukup

Bell Laboratories
Murray Hill, New Jersey   07974

1. INTRODUCTION

Integrated Circuits are becoming increasingly complex, but competitive pressures require that design and layout time remain roughly constant. As a result various parts of the chip are laid out simultaneously by multiple teams of designers and the parts are then assembled. A basic problem with this approach is that much critical information about chip area and long distance wiring capacitance is not known until after assembly. This means that the designers of the individual parts have little knowledge of loading on their circuits or optimum shape and are unlikely to produce optimum designs. A further problem is that the chip can have state of the art performance only if the layout can be updated quickly to new design rules. This is easy if design rules all shrink proportionately, but actually some design rules may shrink substantially while others are unchanged. This paper will describe a Top Down design technique used on a 100,000 transistor 32-bit CPU which provides accurate information about interconnect wiring and optimum use of chip area right from the start of the design process. In addition the layout can be updated to new, nonproportional design rules automatically to take advantage of state of the art processing.

2. DESIGN METHODOLOGY

(a) Hierarchical design

The design technique requires a parallel structure for the functional hierarchy and the layout hierarchy. Figure 1 shows the functional structure of the CPU. There are four major blocks: the execution unit (RALU), the control Programmed Logic Arrays (PLA's), the random control logic, and the I/O. These are further divided into subblocks.

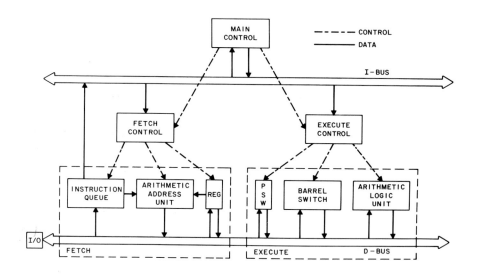

Fig. 1. *Functional Structure of the CPU.*

For example the RALU contains an arithmetic unit, registers, a barrel switch and other structures. At the start of the design process, estimates are made for the amount of logic and the size of each block. At the same time a functional description of each block is made along with a list of all interblock connections. With the size estimates and interconnect information an interblock layout can be made using LTX,[1] an automatic wiring program. An example of the interblock layout is shown in FIG. 2.

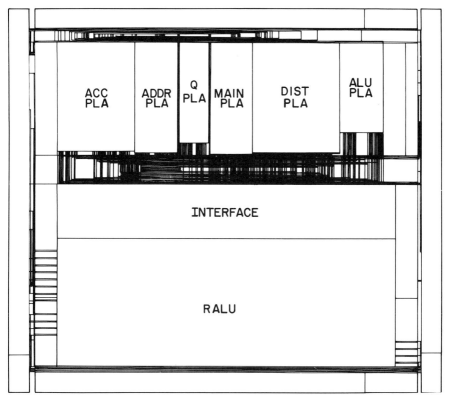

Fig. 2. *Interblock layout of the CPU.*

This layout gives information about capacitance of interblock wires and the area required for them that is normally available in detail only at the end of the layout process. At this point a small effort can change the blocks to reduce wiring area. Now layout and design on each block can proceed in parallel with each block being optimized consistent with known information on loading and interconnect.

An example of a block layout is the RALU which contains 20,000 transistors and is shown in Fig. 3. Here again the layout reflects the functional requirements. A 32-bit databus runs horizontally through each subblock, while control signals run vertically through each bit. This is the most compact way to assure that each bit can be controlled by each control line and makes it easy to divide the layout and design work up among several parallel effects. A uniform bit pitch throughout the RALU insured all parts would fit together when finished. Substantial circuit optimization took place without any significant impact on the overall RALU block connections. This confirmed the utility of the top down approach in predicting interconnect wiring before the content of the blocks has been designed.

Fig. 3. Layout of the RALU part of the CPU.

As blocks are designed in detail, the transistor sizes, capacitance and logic are simulated to verify functionality and speed. Deficiencies are corrected as they are found. For example, the add time of the 32-bit ALU was speeded up from 170 nSec to 120 nSec during the design process without changing its terminal arrangement or block size. Final verification includes simulation of the entire chip at the transistor level using parameters generated automatically from the layout. Since each piece is designed to work with the actual loads that are encountered this should proceed rapidly. In addition use of a simulation based entirely on automatically generated layout data insures a rapid debug of the chip.

(b) *Updatability to new design rules*

Updatability to new design rules is another vital concern for high performance chips. The transition from one generation of design rules to another does not result in a uniform shrink. Each part of the 32-bit CPU is laid out with an updatable technique called gate matrix[2]. This style has been used for all except a few specialized structures, such as PLA and I/O. Using it permits automatic updating because gate matrix symbolic code shows only relative position of structures. All details concerning sizes of gates, contact holes, line widths, etc. are contained in a separate technology file. This file is the only thing which must be changed to reflect new design rules. Then all layouts can be regenerated automatically using unchanged symbolic code. Figure 4 shows the end results of this process for a part of the RALU. Two nonproportional sets of design rules were used to generate these two layouts from the same symbolic code. Previous methods would have required a complete relayout of the circuit to take full advantage of the new design rules.

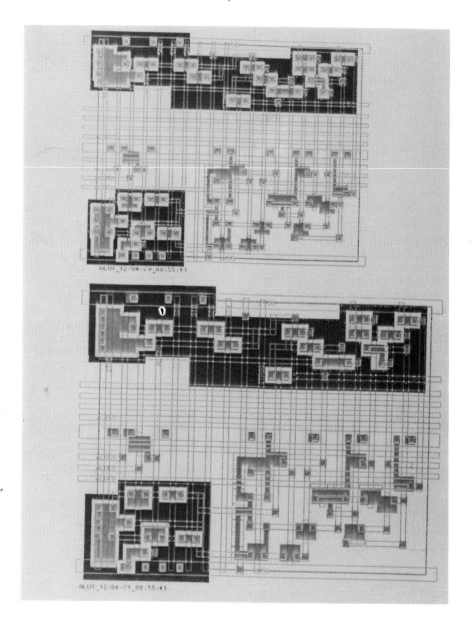

Fig. 4. Layout of a part of the RALU using two nonproportional sets of design rules.

3. CONCLUSIONS

The top down method described here provides a mechanism to obtain detailed and accurate information on area and wiring capacitance early enough in the design process to permit timely optimization of all subblocks on the chip while they are being worked on simultaneously. A speedy debug is assured because a complete transistor level verification of the layout is made. Finally, the entire layout can be updated to new design rules automatically. This permits taking full advantage of the latest design rules.

References

1. Persky, G., et al., "LTX A System for Directed Automatic Design of LSI Circuits," *Proc. 13th Annual Design Automation Workshop*, San Francisco, pp. 399-407 (1976).

2. Lopez, A. D. and Law, H. F. S., "A Dense Gate Matrix Method for MOS VLSI," *IEEE Transactions on Electron Devices*, 27, pp. 1671-1675 (1980).

SYNTHESIS AND CONTROL OF SIGNAL PROCESSING ARCHITECTURES BASED ON ROTATIONS

Hassan M. Ahmed and Martin Morf

Information Systems Laboratory
Stanford University
Stanford, CA 94305
U.S.A

ABSTRACT

A microprogram control strategy is presented for a dual processor speech and signal processing chip which utilizes the CORDIC algorithms. The fundamental nature of plane- and J-rotations to some signal processing algorithms is shown by example. The implementation of these algorithms using vector rotations as primitive operations is then given.

I. INTRODUCTION

Recently, much attention has been afforded to speech and signal processing algorithms and architectures. Such applications are quite computationally intensive, resulting in architectures employing fast multiply and accumulate circuitry [BODDIE at al., 1980], [WIGGINS et al., 1978], [HAMILTON, 1980]. However, many algorithms appearing to possess additions and multiplications as fundamental operations are actually more naturally formulated in terms of elementary plane and hyperbolic rotations (i.e. orthogonal and J-orthogonal transformations). The authors have participated in the architectural definition of a VLSI chip [AHMED et al., 1981a] which performs 2×2 rotations very efficiently using the CORDIC (for COordinate Rotation DIgital Computer) algorithms [VOLDER, 1959]. Although the architecture was aimed originally at the general speech analysis problem, it is in fact ideally adapted to a variety of general signal processing needs.

In this paper, a microprogram control strategy for the chip is described and its application to a variety of signal processing tasks is considered. These examples are intended to demonstrate the fundamental nature of rotations in signal processing and suggest a design methodology for DSP architectures which involves casting the equations to be implemented into a rotation framework which describe the realization.

Research supported by DARPA contracts MDA903-80-C-0331 and MDA903-79-C-0680 and an NSERC of Canada scholarship.

II. CONTROL STRATEGY

The chip descibed in [AHMED et al., 1981a] consists of two CORDIC processors, some scratchpad memory, separate input and output ports (for data flow applications) and a microprogram controller (see figure 1).

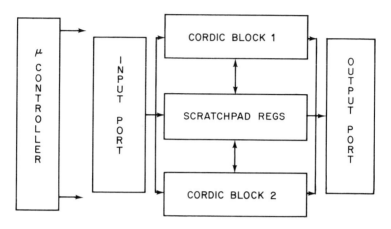

Figure 1: Signal Processing Chip Architecture

A good microprogram control strategy should have a simple structure, be easily programmed and be capable of efficiently implementing the basic operations required by the signal processing algorithms of interest. In this presentation, the controller will be described (based on the CORDIC operations being of interest) and then implementations for a variety of signal processing algorithms will be given.

Simplicity of structure and programming is realized with a two level control philosophy. The higher level or "macrolevel" of operation consists of a set of powerful instructions (figure 2), few in number, which define the functional operation of the chip, e.g. as a speech synthesizer, adaptive equalizer, filter etc. A chip user need only be concerned with this level of operation. Macrolevel instructions may invoke one or more operations at the microlevel, the second level of control. Note that while the prefix "macro" or "micro" is used to distinguish between instructions at the two levels of operation, they both form a part of the controller's microprogram. A microinstruction is a single iteration of the CORDIC recursions (the precise definition depends on which of the four implementations in [AHMED et al., 1981a] is used).

The macrolevel instructions are not uniform in their execution time. As mentioned in [AHMED et al., 1981b], the CORDIC operations are considerably slower than the data transfer or SADD and SSUB instructions. While the latter are said to require one microcycle, the former consume one macrocycle which consists of many microcycles (the exact number depends once again on which of the four realizations is chosen). In fact, the CORDIC instructions such as MUL, JROT etc. are calls to subprocedures which implement the recursive CORDIC algorithms as a sequence of microlevel instructions. In order to avoid processor waiting, as well as in the interest of programming ease, macroinstructions may only invoke operations of the same group (figure 2) in the two processors.

INSTRUCTION MNEMONIC	OPERANDS	OPERATION	COMMENTS
MOVE[1]	src, dst	data xfer	src, dst are CORDIC X,Y,Z regs, scratchpad or I/O
SADD[1]	k	$X+2^{-k}Y$	k is a four bit, unsigned integer
SSUB[1]	k	$X-2^{-k}Y$	
DO[1]	n	initiates loop. $0 \leq n \leq 254$. Do 'forever' if n=255.	
MUL[2]	–	$Y+XZ$	Multiply and Accumulate
DIV[2]	–	$Z+Y/X$	Divide and Accumulate
ATAN[2]	–	$Z+\tan^{-1}Y/X$	
CROT[2]	–	plane rotation of $[X\ Y]^T$ by angle Z	
ATANH[2]	–	$Z+\tanh^{-1}Y/X$	
JROT[2]	–	hyperbolic rotation of $[X\ Y]^T$ by Z	

[1] – denotes GROUP 1 instruction   [2] – denotes GROUP 2

Figure 2: Macrolevel Instruction Set

A simple microcontroller structure is shown in figure 3. A two port memory provides the neccessary microcode to each processor. Separate program counters control program execution at the two levels while a field of the instruction is used for address sequencing via the next address logic (NAL). All of the neccessary control signals to direct the operation of the CORDIC processors as well as the I/O and scratchpad communications are provided by various fields of the microcode. An iteration counter whose sequencing is controlled by the NAL, is provided for simple looping constructs. A loop is initiated with a DO instruction specifying the beginning of a construct. The final instruction of the loop body is signified by a microprogrammed (nanoprogrammed?!) LOOP bit. At this point, control returns to the address following the DO instruction and the iteration counter is decremented. Notice that a *DO forever* facility is provided by setting $n=255$.

Both of the macrolevel processes are tightly coupled in their address sequencing by virtue of a single program counter, PC0. The two port memory appears as a wide single port device at this level of operation. Furthermore, since both processes must be from the same group, the chip may be viewed as an SIMD machine invoking two concurrent operations, with the wide memory output being the instruction (alternately, it can be viewed as a constrained MIMD structure). When a group 2 macroinstruction is executed, program control is transferred to PC1 and PC2 at the microlevel of operation. Now, a true MIMD structure exists with a separate instruction for each processor. These are of course, the actual iterations of the CORDIC algorithms. Microlevel procedures must be of the same length (again for simplicity) and program control is returned to PC0 simultaneously for both processors. It is worth noting that it is frequently not possible to achieve equal length procedures for the various CORDIC instructions due to the need for scaling cycles and pre-rotations [HAVILAND et al., 1980]. However, through a judicious choice of the rotation sequence of the algorithms, efficient procedures of equal length may be obtained as described in [AHMED, 1982].

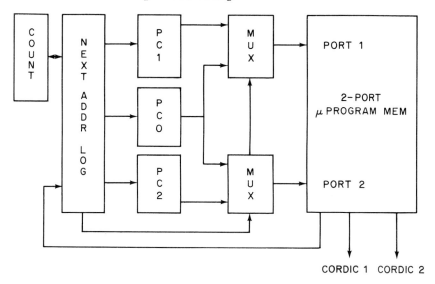

Figure 3: Microcontroller Architecture

Before applying the instructions described above to actual signal processing algorithms, it merits noting that a decision making facility as well as an unconstrained MIMD control philosophy at both levels of operation are being studied.

III. ROTATIONS AND SIGNAL PROCESSING

Efficient single, multi and array processor architectures for many signal processing algorithms may be synthesized based on rotations, since the latter are frequently closely linked with the theoretical basis of the algorithm. Problems of apparently large computational complexity based on multiplications being the primitive operations, become manageable when they are analyzed based on plane- or J-rotations, as has been shown for the linear predictive speech analysis problem [AHMED et al., 1981a] as well as for some matrix factorizations [AHMED-MORF, 1981].

This idea is further demonstrated in the sub-sections to follow, for some simple but very common signal processing applications. The derived processor structures are variations on the ideas of the chip of [AHMED et al., 1981a], as well as the control structure of section II.

IIIa: Speech Synthesis

Many applications involve the synthesis of stored speech segments e.g. consumer products such as SPEAK&SPELL. The synthesis problem is also readily implemented in ladder form using a single processor variant of this chip (figure 4) because each stage of the filter is just a rotation by an amount related to the reflection coefficient of the stage. (for a discussion of the problem, refer to [WIGGINS et al., 1978]).

When the synthesis application involves stored rather than arbitrary speech segments (as would be the case in digital telephony), the implementation can be significantly simplified by storing ϑ_n rather than the reflection coefficient, ρ_n. Even in the telephony case, ϑ_n appears to be a more natural parameter to transmit.

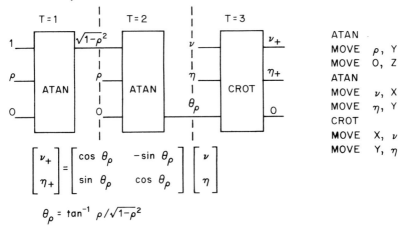

Figure 4: Ladder Form Speech Synthesizer

IIIb. Adaptive Equalization

An important element of today's high speed modems is the adaptive equalizer [LUCKY et al., 1968] which compensates for the lowpass filter characteristics of the channel. The most common realization of this device (figure 5) is a complex transversal filter whose coefficients are adjusted using the least mean square (LMS) gradient algorithm [WIDROW, 1970] and is described by:

$$z_k = \sum_n c_n^t r_{k-n} \qquad (1)$$

$$c_n^{t+1} = c_n^t - \Delta e_n. \tag{2}$$

where

c_n^t is the nth complex tap coefficient at time t
r_n is a complex input sample (applies to all linear modulation schemes)
z_k is the complex equalizer output
Δ is a real adaptation constant
e_n is a complex error signal supplied from elsewhere in the modem (referred to as decision feedback equalization)

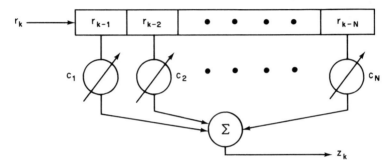

Figure 5: LMS Adaptive Equalizer

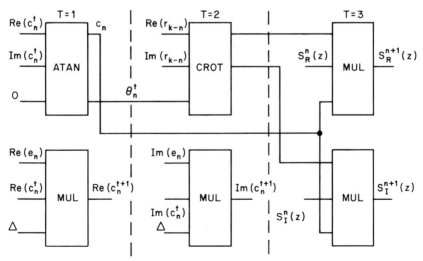

$S_R^n(z)$, $S_I^n(z)$ are real and imag. partial sums of z_k at order n.

Figure 6: Adaptive Equalizer Implementation

The equalizer is a significant portion of the modem data processing, especially when the complex multiplications are realized as real multiply and accumulate operations. Again, this problem is more naturally represented in terms of rotations since these are a well known representation of complex multiplications. An implementation of one equalizer iteration is given in figure 6.

Notice that the modem prefiltering can also be done as defined by (1) except that the filter is time invariant so that it is desirable to store ϑ_n rather than the filter coefficients themselves. By the same token, it is more convenient for the adaptive equalizer to update ϑ_n rather than c_n in (2) in order to simplify the implementation of figure 6. For small perturbations (e.g. after equalizer training), (2) may be simplified to:

$$\vartheta_n^{t+1} = \vartheta_n - \frac{\Delta \operatorname{Im}(e_n)}{\operatorname{Re}(c_n)}, \qquad (3)$$

where

$\operatorname{Im}(e_n)$ = imaginary part of e_n
$\operatorname{Re}(c_n)$ = real part of c_n.

Both the LMS equalizer and the speech synthesizer are naturally described by rotations. Efficient implementations based on the rotation framework can be obtained either with the present chip or uniprocessors which are variations of the present architecture and control structure. Finally note that there has been considerable interest recently in exact least squares ladder forms applied to equalizers [SHENSA, 1980]. Their implementation is identical to the speech analysis problem described in [AHMED et al., 1981a].

IIIc. Array Signal Processors

With the rapid increase of active circuit density on VLSI circuits, much attention has been devoted to array processor architectures for large signal processing applications which require matrix manipulations e.g. [KUNG, 1979]. Array architectures based on fast multiply and add primitives have been offered as solutions to some problems in estimation, image formation and beamforming which involve matrix factorizations (Gaussian elimination, Cholesky factorization, orthogonal decomposition for example).

Once again, it is possible to show that rotations are more fundamental primitives in these matrix manipulations and array architectures employing processors of the type discussed here are naturally derived. A pipelined array implementation of the fast Cholesky algorithm [MORF, 1974] is shown in figure 7. Due to lack of space, details of the algorithm must be deferred to the conference presentation or to [AHMED-MORF, 1981].

The array of figure 7 is triangular, being $n-1$ processors *high* and n processors *wide*, where n is the dimension of the Toeplitz matrix to be factored. The X and Y inputs to the bottom row correspond to the first column of the matrix and the same column with its first element zeroed as described in [AHMED-MORF, 1981]. Many matrices are fed to the array in sequence to exploit its pipelined nature. Factorizations are completed at a rate of $2n-3$ timesteps, with the columns of the factors appearing at the outputs of the processors. Angles for the J- orthogonal transformations of the algorithm are computed in the leftmost processors of the array and propagated one processor to the right at each time step. Such local communication is conducive to rapid VLSI implementation since the interconnect is localized. Temporal computation wavefronts propogate through the array with slope 1/2 and it is this slope which determines the multiplicative constant in the completion rate of the factorizations given above.

It is important to note that the final calculation in the factorization has been oversimplified in figure 7. This is a special case which requires feedback of ϑ_{n-1} into the processor where it was computed to calculate the final element of the Cholesky factor.

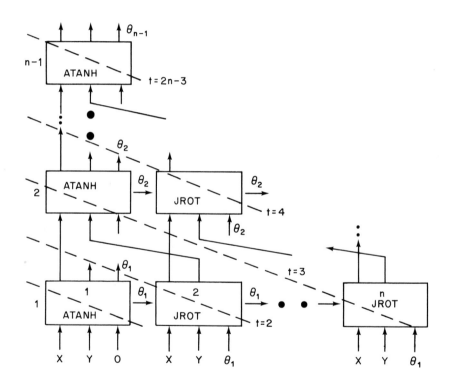

Figure 7: Array Implementation of Fast Cholesky Algorithm

IV. CONCLUSIONS

A two level microprogram control strategy for a dual processor signal processing chip has been presented which is easily programmed. Vector rotations have been shown to be more effective primitive operations than multiplications in a variety of signal processing applications. When the equations are cast into a rotation framework, they are readily implemented in a machine structure equivalent or similar to the present dual processor chip. Therefore, in a sense, the dual processor chip represents a canonical or core signal processing architecture.

Future research involves enhancing the controller with a decision making facility, complete MIMD operation as well as a multiple bidirectional I/O capability for interprocessor communication in array structures.

BIBLIOGRAPHY

[AHMED, 1982]
"Convergence Properties of CORDIC Algorithms," in preparation.

[AHMED et al., 1981a]
"A VLSI Speech Analysis Chip Set Based on Square-Root Normalized Ladder Forms," *Proc. 1981 ICASSP*, Atlanta, GA, March 30, 1981.

[AHMED et al., 1981b]
"A VLSI Chip Set Utilizing Co-ordinate Rotation Arithmetic," *Proc. of 1981 Intl. Symp. on Circuits and Systems*, Chicago, Illinois.

[AHMED-MORF, 1981]
"VLSI Array Architectures for Matrix Factorization," *Proc. of 1981 Workshop on Fast Algorithms for Linear Systems*, Aussois, France.

[BODDIE et al., 1980]
"A Digital Signal Processor For Telecommunications Applications," *Proc. of 1980 Intl. Solid State Circuits Conf.*, San Francisco, California.

[HAMILTON, 1980]
"Speech Processor on Single Chip Talks at Low Bit Rate With Novel Coding Technique," in *Electronics Review section of Electronics*, Vol. 53, No. 24, November 6, 1980

[HAVILAND et al., 1980]
"A CORDIC Arithmetic Processor Chip," *IEEE Trans. on Computers*, Vol. C-29, No. 2, February, 1980.

[KUNG, 1979]
"Let's Design Algorithms for VLSI Systems," *Proc. of 1st Caltech VLSI Symposium*, pp. 65-90, January 1979.

[LEE-MORF, 1980]
"Recursive Square-Root Ladder Estimation Algorithms," *Proc. 1980 ICASSP*, Denver, CO, April 9-11, 1980,

[LUCKY et al., 1968]
Principles of Data Communications McGraw Hill, 1968.

[MORF, 1974]
"Fast Algorithms for Multivariable Systems," *Ph.D. Thesis* , Dept. of Electrical Engineering, Stanford University.

[SHENSA, 1980]
"A Least-Squares Lattice Decision Feedback Equalizer," *Proc. of 1980 Intl. Conf. on Communications*, Seattle, Washington.

[VOLDER, 1959]
"The CORDIC Trigonometric Computing Technique", *IRE Trans. on Electronic Computers*, Vol. EC-8, No. 3, pp. 330-334, Sept. 1959.

[WIDROW, 1970]
"Adaptive Filters," *in Aspects of Network and Systems Theory* (Kalman, DeClaris), Holt, Rinehart and Winston, 1970.

[WIGGINS et al., 1978]
"Three-Chip System Synthesizes Human Speech," *Electronics*, Vol. 51, No. 18, August 31, 1978

MOSAIC: A MODULAR ARCHITECTURE FOR VLSI
SYSTEM CIRCUITS*

J. Alves Marques

INESC
Rua Alves Redol, 9
1100 Lisboa, PORTUGAL

1. INTRODUCTION

The system circuit will be an important percentage of the VLSI market. The MOSAIC project is mainly aimed at developing a design methodology for this very complex circuits (more than 100 000 MOS).
Within the scope of this project a number of design steps has been specified and the corresponding software tools defined.
In the present paper it is essentially described the distributed architecture for implementing this special class of circuits, which is a fundamental part of the MOSAIC methodology.

2. SYSTEM CIRCUITS

By system circuit we mean the integration of a complete computational system, including processing capability, memory, I/O interfaces, in a single chip. As potential market for this class of circuits we see the replacement of small microcomputer systems in applications where it is essential to minimize the number of chips used, the assembling and maintenance costs and to improve the reliability (e.g.: small telecommunication systems, peripheral controllers, machine tool controllers, consumer applications, etc.).
An important characteristic of these small systems is that they do not require a considerable computational power (in the sense of the amount of mathematical calculations involved) but must execute several specialized I/O functions

---

* *This work was carried out while the author was staying with the Computer Architecture Team of IMAG - France*

which require dedicated hardware interfaces, for handling high I/O rates, analog functions, etc.

3. PROBLEMS RAISED BY VLSI

Adaptation of the internal structure of the circuit

The implementation of a system in a single chip can be better understood if we first review the way systems were designed to meet a specific application.

At the SSI/MSI a system was formed by assembling several logic circuits according to a logic design. For LSI technology, programming was introduced as a method for adapting a very complex hardware to the specifications of a given application.

For system circuits a fully programmable solution (for instance, a single-chip microcomputer) can not be used because some functions must be implemented by means of specialized hardware structures (e.g. analog I/O functions, communication protocol controller, a CRT controller, etc..

At this level of complexity, exclusive use of programming is not possible. It is therefore necessary to include hardware blocks optimized both in area and performance.

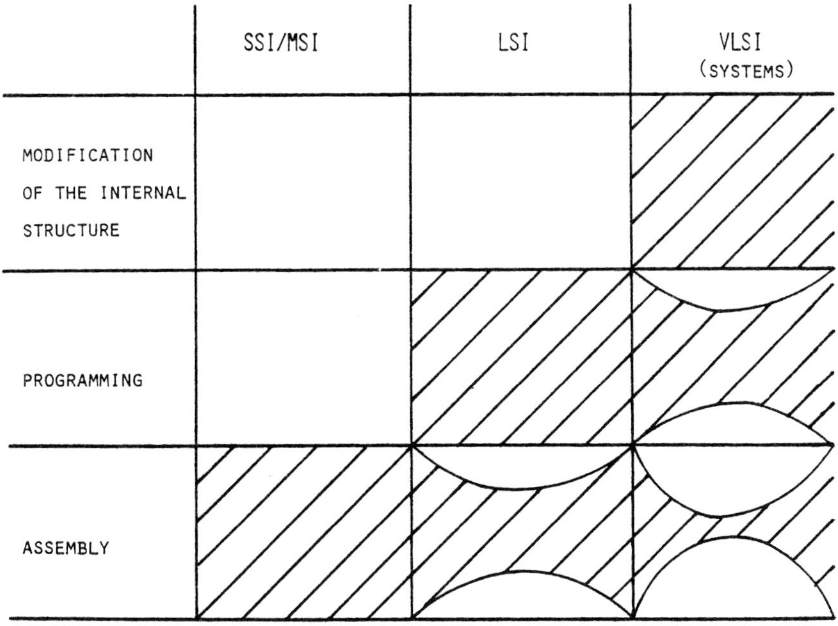

Fig. 1 Adaptation of technology to an application

The most important problem that was raised when establishing the methodology was how to use specialized hardware blocks without being forced either to custom design the whole circuit (what would become impossible due to the exponential rise of design time with complexity |1|) or to use automatic routing algorithms which lead to an inefficient use of the silicon area.

Asynchronism in VLSI

The VLSI circuits of the future can not have a complete synchronous behaviour because the delays introduced by the lines which distribute the clock inside the chip will be of the same order of magnitude of the switching times of the elementary gates |2|.

Beyond a certain limit it will be no longer possible to consider that the clock pulses are received by all the internal elements at approximately the same time. As there is no forseen technological solution, this problem must be overcome by using an internal architecture adapted to the asynchronic behaviour of the circuit.

4. MOSAIC ARCHITECTURE

In MOSAIC a solution to these problems is proposed consisting in the design of a system circuit by using, as building blocks, *functional processors* assembled around a standard communication Kernel |3|.

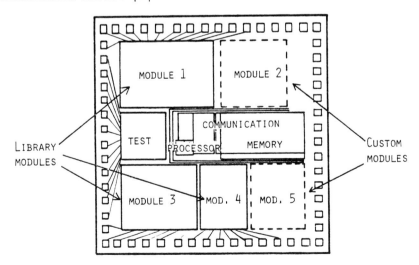

Fig.2 Circuit topology

For a given application, the designer either makes use of a set of defined and tested functional processors kept in a library or may develop customized processors adapted to some new specific need.

The functional processors, designated as modules in this work, have a complexity of 5000 to 10000 transistors (about the same of the existing peripheral circuits of microprocessor systems) and will be connected to the communication system through a standard interface.

All the actions inside the module are synchonous. The modules have a relatively small area and the clock line delays are a function of distance through a diffusion equation.

5. COMMUNICATION MECHANISM

Bus level

In order to keep the modularity of the architecture and to reduce the number of wires inside the circuit we use a single bus to which all the modules are connected through a *standard interface*. By partitioning the circuit in isochronic zones the problems related to the clock skew are transfered to the communication mechanism. A possible solution would be to use a completely asynchronous protocol |4|. However this protocol would imply the use of complex hardware and would be difficult to specify and test.

We prefer to use a simpler solution by considering an hierarchy of isochronous zones inside the circuit.

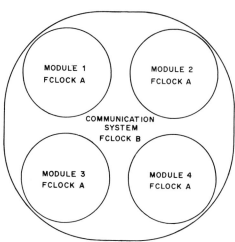

Fig 3   Hierarchy of isochronic zones

The influence of the delays in the clock lines are proportional to the clock period. Therefore the use of large clock periods wherever possible helps to diminish the influence of those delays.

A fast clock is sent to the modules, the clock signals received by different modules will have different phases due to unequal delays. This clock is divided by an appropriate factor (depending on the area of the circuit) for generating a slower frequency which controls the actions of the communication protocol executed by the interfaces.

Since the communication between the interfaces and the modules is asynchronous a special circuit based on a PLL was designed for equalising locally the phase of the execution and communication clocks, thus reducing the probability of occurence of a metastable state in the memory elements of the interface.

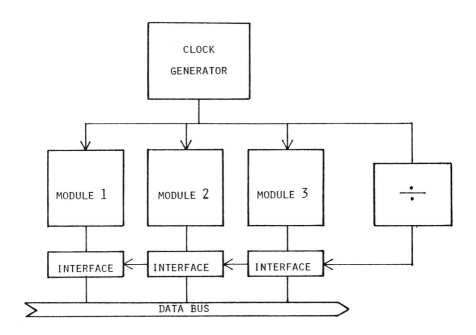

Fig. 4 Logical distribution of the clocks

System level

A distributed solution using a network like protocol is conceptually very promising. However the evaluation of the hardware mechanism required for implementing this protocol has shown that the area occupied by these interfaces would be too large at least for the technologies available in a near future.

A common memory can not be used because it would become an important bottleneck to the system performance specially if the limited transfer rate of the communication bus is considered.

In order to avoid this bottleneck while keeping the area within reasonable limits an intelligent memory is used as Kernel of the communication system.

The intelligent memory executes the algorithm and the synchronisation associated with the communication variables that have been defined by us as an hardware implementation of the *monitor concept* of Hoare and Hansen |5|.

This solution overcomes the problem of the limited bandwith of the communication bus. The procedures of the monitors are executed in the intelligent memory thus avoiding the transfer of the code and of the capabilities of the monitor between the memory and the functional processors. Since all the elements that use the bus have a certain degree of intelligence, the control information has a code format thus reducing the number of data transfers.

The intelligent memory consist of a specialized microprogrammed processor which executes the algorithms of the monitors and of a dynamic memory.

In spite of a relatively short execution time, the communication Kernel may bottleneck the system in applications where the modules are called to execute real time I/O processes with high access rates. In order to avoid the congestion of the intelligent memory, monitors with high access rates may be descentralized and implemented locally close to the modules that use them more often. The local implementation of a monitor consists of a hardware finite state machine (PLA with feedback lines) which executes the algorithms of the monitor and a small local memory.

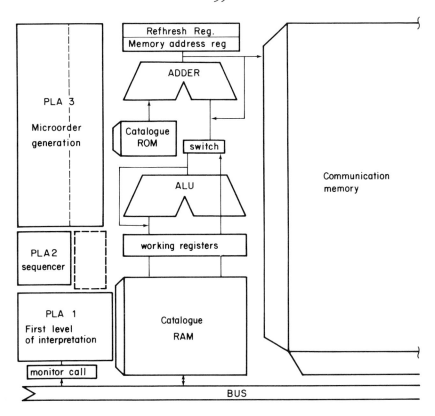

Fig. 5   Communication Kernel

6. DESIGN METHODOLOGY

The efficient use of specialized processors, with respect to design time and the silicon area occupied, was the goal of the design methodology.

The method for designing a new circuit is divided in two major phases. Ih the first one the application is analysed as a whole and the system is specified and evaluated. The most important steps of this phase may be summarized as follows:

- Specification of the functional behaviour of the system, using a high-level language inspired in concurrent Pascal |7| and Modula |8|, which allows a functional simulation.

- Evaluation of system performance using a performance analysis program to simulate the behaviour of the circuit, taking into account the constraints imposed by the real time I/O functions. Choice of the implementation of the communication variables (intelligent memory or distributed) and definition of the performance requirements for the new modules.

- The microprogramming of the Kernel processor and the personalization of the PLA's for the distributed state machines are automatically generated by a compilation of the high level language specification of the monitors.
- A global area evaluator produces a floorplane of the circuit representing the areas occupied by the communication Kernel, the library modules used and the area estimated for the new modules required by a specific application.

At the end of this global analysis an iterative process will allow a compromise between the initial specifications, the area and the performance obtainable from a specific technology.

In the second part of the method the modules that do not exist in the library are designed and laied out according to the results obtained by the previous analysis.

The design of a new module starts from a clear specification of its functions, the performance requirements and an estimate of the area. In the layout of the modules some topological constraints must be respected in order to insure a smooth adaptation to the global structure.

As illustrated by this summary of MOSAIC, we did not attempt to create a general method of CAD for VLSI |9,10| but rather to propose a design methodology for the implementation of one chip systems based upon a global architecture which was optimized from the start, in the execution of this particular class of applications.

ACKNOWLEDEGMENT

I which to thank M. Lança and J.L. Fernandes for reviewing a draft of this paper.

REFERENCES

1. Latin B., "VLSI Design Methodology the Problem of the 80's for Microprocessors", 16th DAC 1979.
2. Mead and Conway, "Introduction to VLSI System" (chap. 7), In Addison Wesley, 1979.
3. Alves Marques, J. "Mosaic: A Design Methodology for VLSI Custom Circuits", 6th ESSCIRC, 1980.
4. Steiz, C., "Ideas About Arbiters", In Lambda nº1, 1980.
5. Hoare, C.A.R., "Monitors an O.S. Structure Concept", CACM Octobre 1974.
6. Alves Marques, J. "A Multiprocessor Architecture Adapted to VLSI Custom Design", Euromicro symposium, London 1980.
7. Brinch Hansen, P., "The Programming Language Concurrent Pascal", IEEE Trans. on Software Engineering, June 75,

8   Wirth, N., "Modula, a Programming Language for Multi-programming", Software Practise and Experience, January 1977.
9   Preas, B.T. and Gwyn, C.N., "Methods for Hierarchical Automatic Layout of Custom LSI Circuits Masks", 15th DAC, 1978
10 Oakes, M.F., "The Complete VLSI Design System", 16th DAC, 1979.

SESSION 2

THE VLSI CHALLENGE: COMPLEXITY BRIDLING

Martin Rem

Department of Mathematics, Eindhoven University of Technology, P.O. Box 513, 5600 MB Eindhoven, The Netherlands

1. INTRODUCTION

VLSI is a new medium for the realization of computations. It is a medium with amazing properties, properties that will profoundly affect computer science. That profound effect is not caused by the fact that VLSI allows us to make microprocessors. Microprocessors are just processors as we have known them for more than twenty years. They may —and probably will— have a considerable social effect, but they hardly exploit VLSI's amazing properties. The fundamental property of VLSI is that it is a medium in which computations can be realized that exhibit an unrivaled degree of concurrency. We are not referring to a few cooperating processes but to a surface of thousands, and in the future possibly millions, of simultaneously active computing elements.

If VLSI can be used as an ultraconcurrent realization medium for computations, how then do we design such computations? How do we design computations that permit ultraconcurrent realizations? We know how difficult programming for sequential machines can be. The design of programs that are to be realized as VLSI circuits will be much more difficult. This difficulty is primarily caused by the many mutual dependencies introduced by concurrency, that is, by the sheer complexity that uncontrolled concurrency inflicts on us. We will not be able to design ultraconcurrent computations reliably if we do not learn to bridle their inherent complexity.

The thesis of this paper is that complexity control —a conditio sine qua non— will lead to hierarchical structures. We call the parts that constitute these hierarchies components. The effective reasoning about hierarchical structures requires that we learn how to specify the net effects of components, their "semantics", in a formalism that is independent of the way in which they realize their net effects. Because such specifications are abstractions they will of necessity be of a mathematical nature. Of particular interest are the composition rules, because they are the mechanisms by which the net effect of the composite can be deduced from the net effects of the composing parts.

2. ABSTRACTION AND HIERARCHIES

In order to be able to design large, internally communicating, programs reliably, we must learn how to argue about these structures in a way that prevents uncontrolled growth of the formal reasoning required. The well-known technique to achieve this consists in partitioning —also known as modularizing— our programs into subprograms, or components as we like to call them. A program is a component. It consists of subcomponents and a scheme defining how the subcomponents constitute the component, how they are "connected". Subcomponents are again components. The whole component thus exhibits a tree structure. Such tree-structured components are known as hierarchical components.

Given the specifications of the subcomponents and the way in which the subcomponents constitute the component, we must be able to deduce the meaning of the whole component. The amount of formal reasoning required for this deduction should not increase with the sizes of the subcomponents. Otherwise the method would become impractical for trees of more than a few levels. These larger programs are the very ones for which hierarchical structuring is essential; small programs can be designed reliably with any technique.

The consequence of the observations above is that the specification of a component should *not* reflect its internal structure. It must specify only how the component looks from the outside. It must define its net effect only, that is, it must define the possible communications with its environment. The specification of the possible communications must be reasonably simple, irrespective of whether it concerns a large component (a tree with many levels) or not. Otherwise the component will be too complicated to be useful; or phrased differently, otherwise the amount of formal reasoning will not be constrained effectively.

In specifying the possible behaviours of a component we decided to abstract from its internal structure. Two components with equal specifications may be very different internally, but this knowledge cannot be used when we employ them to build larger components. Their specifications are phrased in terms of their possible communications, and two components with equal specifications are to all intents and purposes equal components. The effect of the composite does not change when one is substituted for the other.

Because the specifications of components are abstractions they will be of a mathematical nature. Notice that we have not chosen them to be abstract for the reason of making things difficult. The specifications are abstractions because we want to keep matters simple. The determination of the effect of a large program would just become too complicated if we had to take the internal structures of the subcomponents into account. We have already said that the specifications should be reasonably simple. Nevertheless, they must never be vague or imprecise. They are all we have and may use. And only if we can fully "trust" them will we not be tempted to "look inside" the subcomponents.

As a general rule the more subcomponents a component has, the more complicated the determination of their combined effect becomes. This is because we have to take into account all possible ways in which the subcomponents can interact. For this reason we must constrain the number of subcomponents per component. In terms of the tree: the number of successor nodes per node (its degree) must be moderate, not only for nodes near the leaves of the tree, but for all nodes. Only by sticking to this rule may we hope to come away with an amount of formal reasoning that is proportional to the number of nodes in the tree. In regular structures, where many nodes are equal, we will even achieve that the amount of formal labour is proportional to the number of *different* nodes in the tree. That is a good property. It makes the length of the correctness proof of a program proportional to the length of the program text.

Now that we have mentioned the words "correctness proof", we would like to emphasize that the hierarchical structuring technique outlined above is not merely some theoretical model for a posteriori verifications of programs. The method is indispensable when designing programs. It is the very tool by which we understand what we are doing. And is not a proof nothing else than the formal reflection of our understanding?

The reader may have wondered whether it is not simpler to exhibit the correct functioning of a component by simulation rather than by formal reasoning. Unfortunately this is, among VLSI designers, a still widely held belief. The belief is wrong: simulation can never can never show that a component will function properly under all possible communications with its environment. Moreover, when simulating we would probably like to simulate the subcomponents separately, so that we know where the error is located if the simulation exhibits one. But the phrase "where the error is located" makes sense only if we have a precise specification of each subcomponent. And if we have these specifications, why then would we not use them to determine by reasoning whether the ensemble of subcomponents achieves the intended effect? Just like "tracers" have become in disuse among programmers, there is no future for simulators.

3. SPECIFICATIONS AND COMPOSITION RULES

We shall specify the possible communications with the environment by listing all possible sequences of communications. Thus the specification of a component is a (possibly infinite) set of finite-length sequences. Such sequences are called traces. There is no reason why we should be afraid of infinite sets. As a matter of fact computer scientists are well-familiar with them. In language theory a language is a set of finite-length sequences, known as sentences. A component defines a set of traces, just like a grammar defines a set of sentences. The elements of the traces are called atoms. Atoms are indivisable: unlike their physical counterparts they are not composed out of smaller entities.

One may question whether it is fair to specify the possible behaviours of a component by means of traces. After all, the atoms in traces are ordered, whereas communications may be concurrent and thus unordered. In a set of traces concurrent events would be "modeled" by having different traces in which these events appear in all possible orders. In other words, concurrency would give rise to "all possible interleavings".

We believe that interleaving is a legitimate and promising method of dealing with concurrency. It is legitimate because we choose the smallest possible grain of interleaving: the atoms. The interleaving of structured entities (traces of more than one atom) must always be defined in terms of the constituting atoms. The often-heard dogma that interleaving and concurrency are essentially different has impeded progress in this area of research too long now. Interleaving is also a promising method. The reason is that it keeps us on solid ground. We know what finite-length sequences are and we know how to treat them mathematically. We may not abandon such a mathematically simple framework lightly.

Sets of traces are discrete objects. The underlying physical world, however, is not. Aren't we fooling ourselves by postulating a discrete frame of reference? No, we are not. And the reason lies at the very heart of computer science. The programmer designs his computing structures in a discrete universe. The task of the electrical engineer is to realize that discrete universe out of, to all intents and purposes, continuous electrical phenomena. He simulates a discrete universe with continuous tools. We know, or have at least strong reasons to believe, Stucki and Cox (1979), that the latter is possible only if we allow the simulations to take an unbounded time.

It will have become clear that the use of sets of traces as specifications is not intended to model the physical properties of computing engines. The specifications are definitions, mathematical definitions, that should be obeyed by the physical realization. A realization that does not obey the definitions is an erroneous realization. Of course the definitions must be chosen in such a way that they are physically realizable. To ensure realizability there will be conditions that the discrete universe must satisfy. This is very similar to Dijkstra's "healthiness conditions" for weakest preconditions. Weakest preconditions are a nonconstructive formalism for defining the meaning of programming constructs. In Dijkstra (1976) Dijkstra gives five (not independent) conditions for weakest preconditions that together guarantee that the constructs thus constrained do not give rise to logical contradictions.

Every component defines a set of traces. What is the set of traces defined by the composition of two components? If the two components are independent, that is, if there is no communication between them, the set of traces of the composite is the set of all interleavings of the traces of the individual components.

Example 1 Let $V = \{ab\}$ and $W = \{ef\}$. The sets V and W each contain one trace of two atoms. The composition, $V + W$, of these two sets is

$\quad \{abef, aebf, aefb, efab, eafb, eabf\}$

Example 2 $\{ab, cd\} + \{ef, gh\}$ contains 24 traces, each of length 4.

Let us now introduce communication between the two sets of traces in Example 2. The two sets are then not independent any more, that is, there should be atoms that occur in both. We achieve this by equating atoms in the two sets. We could, for example, equate b with e and d with g. Atoms thus equated are considered to be the same. To cater to the occurrence of atoms that are common to different sets of traces we change the definition of interleaving:

Let V and W be sets of traces, and let A be the set of atoms common to V and W . Then V + W is defined as the set of all traces that do not contain atoms of A and that are interleavings of traces of V and W , where the concept of interleaving is extended with the following elimination rule: Let t0 and t1 be traces with the same first atom, then t is an interleaving of t0 and t1 if it is an interleaving of their remainders.

Example 3 V = {ab, cd} , W = {ef, gh} , b = e , d = g . We thus have W = {bf, dh} , which yields V + W = {af, ch} . To guide intuition we point out that the above composition can be interpreted as follows. The environment of V chooses between the atoms a and c . V then communicates that choice to W . W answers by f or by h . Input via a into V + W causes it to answer f , input via c will give output via h . The fact that V + W communicates the choice internally has been "eliminated". The component V + W has only af and ch as its traces.

Usually we will like our sets of traces to satisfy certain constraints. Constraints can be imposed to allow a particular kind of physical realization, for example a self-timed realization, Seitz (1980). We would like such constraints to be invariant under composition. This will constrain the sets of traces we may compose even more. As this is not a treatise on trace theory we do not pursue this line of thought any further.

In this section we have learned that communication is the elimination of common atoms in traces and that communication makes the set of traces smaller. Communication thus makes our constructs more deterministic.

4. SPATIAL LAYOUT

In order for a program to be realized as a VLSI circuit it must be laid out as a more or less planar structure. Let us look again at the analogy between programming for sequential machines and VLSI design. No ALGOL compiler would give the programmer a core dump to inquire whether that is how he would like to have his program compiled. The ALGOL programmer does not want to be bothered with the binary encoding of his programs. The VLSI designer is also a programmer. He designs ultraconcurrent programs and he, likewise, should not be confronted with the details of the physical representation. The layout in the plane must be done automatically, by a silicon compiler, without interference or consultation of the programmer.

For some structures the layout task will be simpler than for others. We, therefore, may want to restrict the types of components that may be composed, or we may give the programmer guide lines as to which compositions will result in smaller areas of silicon being used.

The layout of arbitrary graphs is a difficult task. It has been suggested that it is important, or even essential, that the graphs be planar. The planarity of a graph, however, is an obscuring distinction. Graphs do not have to be planar to lay them out as VLSI circuits, because VLSI circuits are not planar. Just like printed circuit boards they may contain crossings.

Fortunately we do not have arbitrary graphs. Our hierarchical components correspond to tree-like graphs. Balanced binary trees map into the plane very well, Rem (1979). Our difficulty is mainly that we do not have balanced binary trees but arbitrary trees. Another complicating factor is that the sizes of the nodes in our tree do not have to be equal. The layout of arbitrary trees with nodes of different areas requires further research. One thing, however, is certain: the layout algorithm must use the fact that we are dealing with graphs of a limited class. Automatic layout of arbitrary graphs is a hopeless task.

We will not consider it to be a defeat of our method if our layout algorithm would require a special fabrication process for the realization of our programs as VLSI circuits. It could, for example, be that we would like our VLSI circuits to have one or two additional layers of metal. We are willing to complicate the fabrication process if that is the price we have to pay for allowing the programmer to abstract from the physical representation of his products.

5. TESTING

For the same reason that we do not believe in simulating, we do not believe in testing for design errors either. With production techniques being what they are it will, however, remain necessary to have our VLSI circuits tested for fabrication errors. Fabrication errors are an evil that is here to stay.

With the continuing growth of our circuits the testing of them will become more and more time consuming. The testing time of arbitrary circuits is an exponential function of their size. We have bridled the complexity of circuit design by imposing a hierarchical structure. But the fact that hierarchical components consist of subcomponents whose net effects are well-specified not only makes their design simpler, it simplifies their testing as well. We can test the subcomponents separately, followed by the testing of their interconnections. The subcomponents are of course tested in the same way. Employing such a hierarchical test method the total testing time of a VLSI circuit will be proportional to its size.

The separate testing of subcomponents can be done with additional pads that are scattered, at the component boundaries, over the silicon surface. If that would consume too large an area we need a mechanism to measure the electrical currents through the "wires" on the chip. Having a self-timed realization may simplify the testing task. For a malfunctioning in a self-timed system exhibits itself by components getting hung on missing acknowledge signals.

6. CONCLUSIONS

Complexity control requires that we design our programs as hierarchically structured components. It forces us to abstract from the representations of subcomponents and to specify precisely the net effects of all subcomponents. We are so fortunate that the realization of our components as VLSI circuits benefits from their hierarchical structure as well. The layout algorithm and the testing procedures will become simpler than they would have been for arbitrary structures.

Ever since programmers started to talk about the "software crisis" have they been concerned with complexity control. VLSI designers have learned little from the experiences programmers have gained over the last two decades. They still use simulation as their primary tool for judging the correctness of their designs.

The best way to control complexity is to avoid it. Avoid elaborate case analysis, and in general: abstract from irrelevant detail. These abstractions will be of a mathematical nature. VLSI design clamors for a mathematical approach.

Recently mathematicians have started to "discover" VLSI. But most of them have kept themselves busy with complexity theory. Complexity theory is not the theory of coping with the complexity of the programming task. It is the study of "optimality" of algorithms, optimal in the sense that their executions require the least time or space. It is an important area of study, but it does not address VLSI's most urgent problem requiring mathematical attention. It would be irresponsible if our mathematically inclined computer scientists ignored the problem of intellectually mastering the design of ultraconcurrent computing structures. Formal methods are required for the reliable programming of sequential machines; this is a fortiori true when one is confronted with ultraconcurrency. Ignoring this fact can only provide evidence of lack of judgment, understanding, and insight in the problems of computer science.

ACKNOWLEDGEMENTS

Jan van de Snepscheut is gratefully acknowledged for his encouragement and his technical contributions. The elimination rule for sets of traces is an example of the latter. The ideas expressed in this paper have benefited from many valuable discussions with Carver A. Mead and Charles L. Seitz.

REFERENCES

Dijkstra, Edsger W. (1976). "A Discipline of Programming". Prentice-Hall, Englewood Cliffs, N.J.
Rem, Martin (1979). Mathematical aspects of VLSI design. *In* "Caltech Conference on VLSI" (Ed. C.L. Seitz). pp. 55-64. Computer Science Department, California Institute of Technology, Pasadena, Cal.
Seitz, Charles L. (1980). System timing. *In* "Introduction to VLSI Systems" (C.A. Mead and L.A. Conway). pp. 218-262. Addison-Wesley, Reading, Mass.
Stucki, M.J. and Cox, Jr., J.R. (1979). Synchronization strategies. *In* "Caltech Conference on VLSI" (Ed. C.L. Seitz). pp. 375-394. Computer Science Department, California Institute of Technology, Pasadena, Cal.

RECOGNIZE REGULAR LANGUAGES WITH PROGRAMMABLE BUILDING-BLOCKS

M. J. Foster and H. T. Kung

*Department of Computer Science
Carnegie-Mellon University
Pittsburgh, Pa. 15213, USA*

1. Introduction

Construction of future VLSI systems will rely on the use of *programmable building-blocks*. A building-block consists of a set of cell designs together with rules for combining the cells into larger circuits, and for using these circuits in large systems. The PLA (programmable logic array) for example, is a programmable building-block frequently used for implementing random logic. Because the structure of a building-block is fixed and prespecified, layout generators, simulators and other high-level design tools can be used effectively. Thus using building-blocks helps manage the complexity of VLSI design. A *programmable* building-block can be "personalized" to realize various functions. If building-blocks are programmable, designers can proceed to high-level designs before all the low-level functions are specified. This often speeds up the design process and increases the flexibility of the final system. These advantages of using programmable building-blocks have already been demonstrated in several recent projects by Holloway *et al.* (1980), and Stritter and Tredennick (1978). This paper proposes a new programmable building-block for constructing efficient circuits that recognize regular languages.

Regular languages are precisely those languages that can be recognized by finite-state machines (see, *e.g.*, Hopcroft and Ullman (1979)). They are well

---

This research was supported in part by the Office of Naval Research under Contract N00014-80-C-0236, NR 048-659, and in part by the Defense Advanced Research Projects Agency under Contract F33615-78-C-1551 (monitored by the Air Force Office of Scientific Research). M. J. Foster was supported in part by a National Science Foundation Graduate Fellowship.

suited for describing identifiers in a programming language or patterns to be matched by a text editor, and for modeling processes associated with electronic circuits and nervous systems. Language recognizers are often used as components in larger systems, such as controllers and sequencers. For example, Haskin (1980) has recently suggested using language recognizers as term matchers in special-purpose database machines. We show that by combining some basic building blocks a recognizer circuit for a language can be constructed automatically from the regular expression describing that language, and a circuit so constructed can itself be a building-block for constructing larger systems.

2. Basic Ideas

To motivate the construction of recognizer circuits using this building block we present several examples of increasing complexity. Our first example is a linear pipeline that can recognize concatenations. Figure 1 shows a pipeline that recognizes any three character pattern. Before the computation starts, the pipeline is loaded with the pattern ABC, one pattern character at each recognizer cell. During the computation the cells are synchronized to operate together on discrete clock ticks, or beats. On each beat, the text to be matched moves through the pipeline from right to left, and the results of the match move from left to right. Data in both streams are separated by one cell to permit each character to meet every result. On each beat, every cell that is active compares its prestored pattern character with the text character received from its right, then sends the text character on to its left. The cell AND's this comparison result with the result received from the left, and sends the new result to the right. The result of comparing a text string with the pattern is available from the pipeline on the beat after the last text character is input, thus a *constant response time* is achieved. Figure 1 traces the action of the pipeline for several beats. Cell contents on each beat are shown underneath the corresponding cells.

Our second example is a tree-structured pipeline that can recognize any regular expression consisting of a union of several concatenations followed by a single concatenation. Figure 2 shows a tree-structured pipeline programmed to recognize the language generated by expression (AB + C)DE. This is an obvious extension of the linear pipeline in Figure 1: characters fan out to both branches of the tree when they reach the " + " node, and results from the two branches go through an OR gate at that node. Neither characters nor results are stored in the " + " node; they just pass right through. Once again, the result of matching a text string against the pattern is available one beat after the last text character goes into the pipe.

This pipeline scheme cannot be extended in the obvious way to expressions such as A(BC + D)E, which contain a union preceded by a concatenation. If we try to use a pipeline like that in Figure 3, where

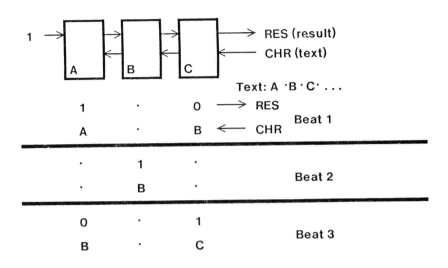

Fig. 1. *Circuit Programmed for the Expression* ABC

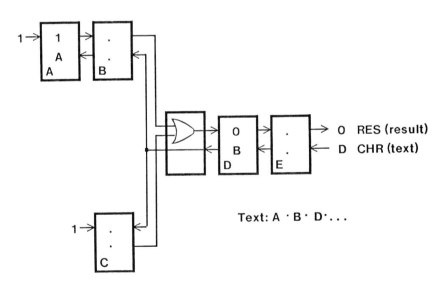

Fig. 2. *Circuit Programmed for the Expression* (AB + C)DE

characters and results flow around both branches of the loop, it is impossible to maintain synchronization when the branches differ in length. Instead we must add a third data stream called the *enable* stream, as shown in Figure 4. On each beat the enable stream moves from right to left. At the end of a branch, the enable is fed into the result stream.

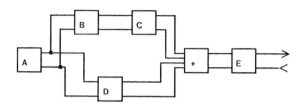

Fig. 3. *Obvious (Wrong) Extension of Tree for* A(BC + D)E

A correct pipeline for the expression A(BC + D)E is shown in Figure 4, together with a record of the contents of each cell for several consecutive beats. The 1's in boxes track a successful match through the pipeline from cell A to cell E. Notice that the text characters are sent through the " + " node to the A node, as well as to the C and D nodes. Thus on each beat, the same text character is in all three of those nodes, and on the next beat it is in the B node. The result from the A character cell fans out (through the " + " node) to the enable stream of both the BC and D branches of the tree. The match results from the A cell thus reach the result stream input to both the B and D cells in synchrony with the character-string.

As is shown in the next section, similar tree-structured recognizers can be built for any regular expression. Notice that, unlike the recognizers of Floyd and Ullman (1980) and Mukhopadhyay (1979), our circuits achieve a constant response time, and do not require broadcast of the text characters. Our recognizers are thus well suited to VLSI implementation, in which broadcast is slow, but local communication is fast.

3. Recognizers for Arbitrary Regular Expressions

How can we construct a recognizer for an arbitrary regular expression? In this section we describe three types of basic cells and give a procedure for hooking them up to form any recognizer.

Three kinds of cells are used in constructing recognizers: comparators for single characters, and combinational cells corresponding to the union (" + ") and Kleene closure ("*") operators. Each of the basic cells of a recognizer has one or more data paths passing through it, with three data streams on each data path. The CHR and ENB streams, which flow from right to left, carry the text characters and enable bits. The RES stream, which flows from left to

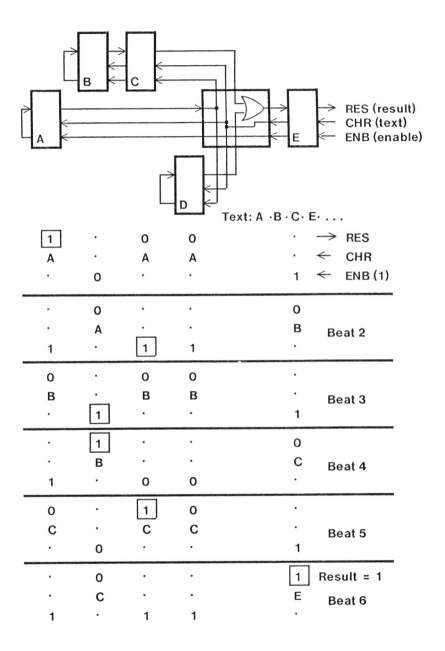

Fig. 4. *Correct Extension of Tree for* A(BC + D)E

right, carries the result bit. We can hook these cells together to form recognizers by connecting the data path at the right side of one cell to a data path at the left side of another cell.

The character comparator is shown in Figure 5, together with a symbol used in designing large recognizers. This cell is similar to the character comparator described by Foster and Kung (1980). On each beat, the cell performs these steps:

1. Compare the text character with the stored pattern character X in the PAT register, AND that result with the RES register, and pass the one bit result to the right.

2. Pass the CHR and ENB registers leftward and receive new contents from the right.

3. Receive new contents for the RES register from the left.

The cells for the " + " and "*" operators contain only combinational logic, and are shown in Figures 6 and 7.

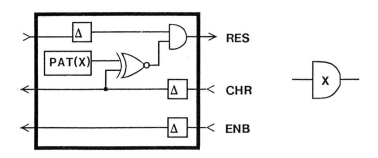

Fig. 5. *Character Comparator Cell*

To construct recognizers for arbitrary regular expressions, these cells must be combined into larger circuits. We use a new technique for combining the cells in which a context-free grammar describes both the structure and function of the final circuit. Terminal symbols in the grammar correspond to basic cells, and semantic actions attached to the productions of the grammar tell how to hook them together.

To construct a recognizer for a regular expression, we parse the expression using the grammar:

R ::= P | RP
P ::= ⟨letter⟩ | (R)* | (R + R)

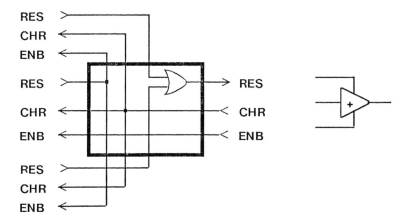

Fig. 6. " + " *Operator Cell*

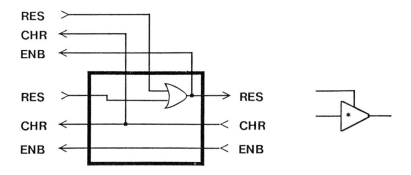

Fig. 7. "*" *Operator Cell*

Each symbol of this grammar represents a kind of circuit. The <letters> represent the character comparator cells, for example, and the non-terminal symbol R represents a recognizer for a regular expression. Each production has an associated semantic rule that tells how to connect the circuits on the right side of the production to form the circuit on the left side. Every time a production is used in parsing the regular expression, its semantic rule is used to add to the circuit.

This syntax-directed construction technique eases verification of functional correctness, and other properties of the resulting circuits. Proof of a single theorem for each production in the grammar will verify the correctness of *any* recognizer. We attach to each symbol in the grammar a predicate describing its circuit. For each production in the grammar, we then prove that if circuits satisfying the predicates on its right hand side are connected according to its semantic rule, then the resulting circuit satisfies the predicate on the left hand side. This verification technique promotes confidence that large recognizers will work as expected.

4. PRA: Programmable Recognizer Array

The pipeline circuits constructed above form ternary trees, so each of them can be laid out in an area efficient manner, as noted by Floyd and Ullman (1980) and Leiserson (1981). This section describes an alternative to the approach of individually laying out the tree corresponding to each recognizer circuit. We propose the use of a single, compact layout, called the PRA (programmable recognizer array), that can be personalized to recognize the language specified by any regular expression. Our methods are similar to those of Leiserson (1981).

For a recognizer of size n, we lay out n basic cells on the bottom line of the array, and provide $O(\log n)$ channels in the top portion of the array for data paths parallel to the line, as shown in Figure 8(a). To configure the layout for a particular tree, we route the edges of the tree through the channels. Ternary trees have a constant separator theorem, so that by removing a single edge of the tree we can split it into two subtrees of roughly equal size. We split the line of cells into two lines, one for each of the subtrees, and use one channel to route the data path corresponding to the removed edge between the two subtrees. We then apply the same procedure recursively to lay out the subtrees on their lines of cells. Figures 8 (b) and (c) show the same PRA "programmed" for two different regular expressions.

A PRA of dimensions n by $O(\log n)$ can be programmed to recognize the language generated by *any* regular expression of length n. For the same problem a PLA implementation, as proposed by Floyd and Ullman (1980), would require n by n area in the worst case. For recognizing languages described by large regular expressions, a number of small PRA's can be combined using the syntactic method described earlier.

5. Conclusions

This paper introduces a new programmable building-block for recognition of regular languages. The building-block can be formed (or "programmed") for any regular expression using a syntax-directed construction method, which also allows easy and mechanical verification of circuit properties. Recognizers built using these building-blocks are efficient pipeline circuits

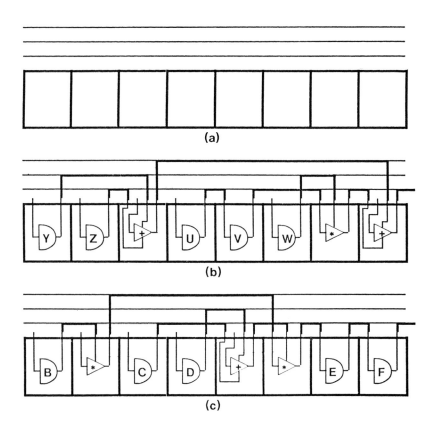

Fig. 8. (a) PRA before "personalization"
(b) PRA for $UVW^* + (Y + Z)$
(c) PRA for $B^*(C + D)^*EF$

that have constant response time, and avoid broadcast as well as long-distance communication. In addition, PRA's provide compact reconfigurable layouts, requiring only $O(n \log n)$ area for regular expressions of length n. Programmable recognizers should be included as one of the building-blocks in the I.C. designer's toolbox.

References

Floyd, R.W. and Ullman, J.D. (1980). The Compilation of Regular Expressions into Integrated Circuits, In "Proceedings of 21st Annual Symposium on Foundations of Computer Science", pp. 260-269. IEEE Computer Society.

Foster, M.J. and Kung, H.T. (1980). The Design of Special-Purpose VLSI Chips, Computer 13(1), 26-40.

Haskin, R. L. (1980). Hardware for Searching Very Large Text Databases, PhD thesis, University of Illinois at Urbana-Champaign.

Holloway, J., Steele, G.L., Sussman, G.J. and Bell, A. (1980). "The SCHEME-79 Chip", Technical Report AI Memo No. 559, EE&CS Integrated Circuit Memo No. 80-6, Massachusetts Institute of Technology Artificial Intelligence Laboratory.

Hopcroft, J.E. and Ullman, J.D. (1979). "Introduction to Automata Theory, Languages, and Computation", Addison-Wesley Publishing Co.

Leiserson, C.E. (1981). "Area-Efficient VLSI Computation", PhD thesis, Carnegie-Mellon University.

Mukhopadhyay, A. (1979). Hardware Algorithms for Nonnumeric Computation, IEEE Transactions on Computers C-28(6), 384-394.

Stritter, S. and Tredennick, N. (1978). Microprogrammed Implementation of a Single Chip Microprocessor. In "Proceedings of the IEEE Eleventh Annual Microprogramming Workshop", IEEE.

A VERY SIMPLE MODEL OF SEQUENTIAL BEHAVIOUR OF nMOS

M. Gordon

*Computer Laboratory, University of Cambridge,
Corn Exchange Street, Cambridge CB2 3QG, UK*

INTRODUCTION

   In this paper we describe a notation for expressing the 'logical semantics' of nMOS circuits. This notation is strongly oriented to the description of sequential behaviour - e.g. behaviour which depends on the storage of charge on electrically isolated gates. It is based on a mathematical model described in Gordon (1981), but we do not go into details of this here. Our approach has been to make everything as simple as possible consistent with being able to represent the essential workings of some typical 'nMOS algorithms'. The most complicated example we have successfully analysed so far is the stack cell and controller in Mead and Conway (1980) (see Gordon (1981) for details). For many kinds of circuits our model is probably far too rudimentary to represent what goes on, nevertheless we feel that our methodology is a good one: by complicating details we hope to be able to model more subtle circuit mechanisms within the same general notational framework. Originally we designed the model for arbitrary register transfer systems such as machine code processors; it is only later that we thought of a way of applying it to circuits. We shall follow this order of development here: first we start by discussing the general principles of our notation, then we specialize it to nMOS.

REGISTER TRANSFER SYSTEMS

   A typical register transfer system is shown in Fig.1. It is built out of three components: a multiplexer *MUX*, a register *REG(n)* (parameterized on its contents n), an incrementer *INC*.

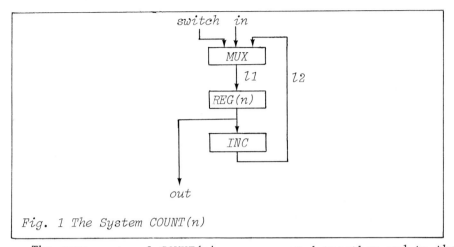

Fig. 1 The System COUNT(n)

The components of COUNT(n) are connected together, and to the outside, by <u>lines</u>. Each line can only carry values of a particular type: line *switch* carries truthvalues (*true* and *false*), all the other lines (namely *in*, $l1$, $l2$ and *out*) carry integers. MUX and INC are <u>combinational</u>; the values on the output lines are a fixed function of the values on the input lines - we abstract away from the finite delay present in real hardware. REG(n), on the other hand, is <u>sequential</u>; the input-to-output function depends on its state n. Thus REG(0), REG(1), REG(2), ... denote the behaviour of the register when it is storing 0, 1, 2, ... respectively. The value on the output line $l1$ of MUX is either the value on the input line *in* if *true* is the value on *switch*, otherwise it is the value on $l2$. The value on the output line $l2$ of INC is one plus the value on its input line *out*. The value on the output line *out* of REG(n) is n. When we 'clock' the register the value on its input line $l1$ is stored; thus if this value is v then clocking REG(n) turns it into REG(v). Suppose we put *true* on line *switch* and 0 on line *in* then 0 will appear on line $l1$ and so if we clock the system 0 will be stored in the register i.e. COUNT(n) will become COUNT(0). If we now put *false* on line *switch* then the value on $l1$ will be the value on $l2$ which will be one plus the value on *out* which is the value (viz. 0) stored. Thus the value on $l1$ will be 1 so if we clock the system again it will become COUNT(1). Clearly successively clocking the system whilst keeping false on line *switch* will cause the system to successively become COUNT(2), COUNT(3), ... We now explain how to express the behaviour of COUNT(n) and its components in our notation.

THE NOTATION

We shall sometimes denote the truthvalues *true* and *false* by
1 and 0 respectively. $\neg, \vee$ and $\wedge$ denote negation, disjunction
and conjunction respectively. If T is a truthvalued expression
and $E1, E2$ are expressions then the value of the expression
$(T \rightarrow E1, E2)$ (pronounced "if T then $E1$ else $E2$") is the value of
$E1$ if T's value is *true*, otherwise it is the value of $E2$. The
value of the expression $E1=E2$ is *true* if $E1$ and $E2$ have identical values, otherwise it is *false*. The <u>behaviour</u> (semantics)
of MUX, $REG(n)$ and INC of Fig.1 are defined in our notation by:

$\quad MUX = \lambda\{switch, in, l2\}.\{l1 = (switch \rightarrow in, l2)\}, MUX$
$\quad REG(n) = \lambda\{l1\}.\{out = n\}, REG(l1)$
$\quad INC = \lambda\{out\}.\{l2 = out+1\}, INC$

Fig.2 Behaviours of the components of COUNT(n).

The general form of a behaviour definition is:

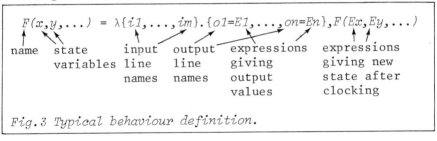

Fig.3 Typical behaviour definition.

This defines a device that outputs the value of Ei on line
oi and when clocked becomes $F(Ex, Ey, \ldots)$ - i.e. the state
changes from $(x,y,\ldots)$ to $(Ex, Ey, \ldots)$. For example, from Fig.2
we see that the value on output line out is n, and when the
register is clocked it becomes $REG(l1)$ - where, in this context, $l1$ denotes the value on the input line. Purely combinational devices like MUX and INC need no state variables as
the values on the output lines are a fixed function of the
values on the inputs.

From the intuitive description of the working of $COUNT(n)$
we gave above we would expect its behaviour to be given by:

$\quad COUNT(n) = \lambda\{switch, in\}.\{out = n\}, COUNT(switch \rightarrow in, n+1)$

Fig.4 Behaviour of COUNT(n).

We now show how to formally manipulate the definitions in
Fig.2 to derive this behaviour. These manipulations constitute

a <u>verification</u> that the behaviour <u>specified</u> in Fig.4 is correctly <u>implemented</u> by the devices in Fig.2.

If $B1,\ldots,Bm$ are expressions denoting behaviours with disjoint sets of output lines (for example, $B1 = MUX$, $B2 = REG(n)$, $B3 = INC$) and if $l1,\ldots,ln$ are line names, then $[\![B1|\ldots|Bm]\!]\backslash l1\ldots ln$ denotes the device obtained by connecting all output lines of $B1,\ldots,Bm$ to input lines with the same name, and also internalizing (or hiding) all output lines with names $l1$ or $l2 \ldots$ or ln. For example, $[\![MUX|REG(n)|INC]\!]\backslash l1\ l2$ denotes the device shown in Fig.1. We shall not give a mathematical model of joining here. Intuitively the idea is that when values are put on the input lines of $[\![B1|\ldots|Bm]\!]\backslash l1\ldots ln$ they propagate instantaneously through the component devices $B1,\ldots,Bm$ to establish values on all the lines according to the connection topology. In real hardware there would be a settling down period — for our model to be valid this must be much shorter than the clock cycle time. When the clock ticks each component device changes state according to the values that have appeared (i.e. stabilized) on its input lines (which may be internal lines of the composite device). For a rigorous description of the meaning of $[\![B1|\ldots|Bm]\!]\backslash l1\ldots ln$ see Gordon (1981).

Manipulating Compositions of Devices

We now describe some manipulative rules that enable us to, for example, show that $COUNT(n)$ (as defined in Fig.4) equals $[\![MUX|REG(n)|INC]\!]\backslash l1l2$ (where MUX, $REG(n)$, INC are defined in Fig.2). First we need some more notation: suppose $E1,\ldots,En,E$ are expressions and $x1,\ldots,xn$ are variables (which may occur in the expressions) then *letrec* $\{x1=E1,\ldots,xn=En\}$ *in* E denotes the value of E when each xi is defined by the (possibly mutually recursive) set of equations: $x1=E1,\ldots,xn=En$. For example, *letrec* $\{x=y+z, y=1+z, z=2\}$ *in* $x+y$ has value 8 because the equations $x=y+z$, $y=z+1$ and $z=2$ define $x=5$, $y=3$ and $z=2$ and so $x+y$ has value $5+3 = 8$. Two rules for manipulating *letrec*-expressions are:

UNFOLDING *letrec* $\{x1=E1,\ldots,xn=En\}$ *in* E
   = *letrec* $\{x1=E1',\ldots,xn=En'\}$ *in* E'
   where Ei', E' are got from Ei, E respectively by replacing occurrences of xi by Ei.

REDUCTION *letrec* $\{x1=E1,\ldots,xn=En\}$ *in* E = E
   if $x1,\ldots,xn$ do not occur in E.

The following example illustrates these rules:

$letrec \ \{x=y+z, y=1+z, z=2\} \ in \ x+y$

$= letrec \ \{x=y+2, y=3, z=2\} \ in \ x+y$ by *UNFOLDING*

$= letrec \ \{x=5, y=3, z=2\} \ in \ x+y$ by *UNFOLDING*

$= letrec \ \{x=5, y=3, z=2\} \ in \ 5+3$ by *UNFOLDING*

$= 5+3 = 8$ by *REDUCTION*

We shall use *letrec*-expressions in behaviour definitions. An example, based on Fig.1, which we motivate later is:

$COUNTIMP(n) = \lambda\{switch, in\}.$
$\qquad letrec \ \{l1=(switch \rightarrow in, l2), out=n, l2=out+1\}$
$\qquad in \ \{out=n\}, COUNTIMP(l1)$

Fig.5.

By several applications of *UNFOLDING* the equation in Fig.5 becomes:

$COUNTIMP(n) = \lambda\{switch, in\}.$
$\qquad letrec \ \{l1=(switch \rightarrow in, n+1), out=n, l2=n+1\}$
$\qquad in \ \{out=n\}, COUNTIMP(switch \rightarrow in, n+1)$

and so by *REDUCTION*:

$COUNTIMP(n) = \lambda\{switch, in\}.\{out=n\}, COUNTIMP(switch \rightarrow in, n+1)$

Fig.6

Notice that the equation for $COUNTIMP(n)$ in Fig.6 is identical to the one for $COUNT(n)$ in Fig.4. The rule *UNIQUENESS* stated below enables us to conclude that $COUNT(n)=COUNTIMP(n)$.

UNIQUENESS If two behaviour expressions satisfy the same equation (of the form shown in Fig.3), then they denote the same behaviour - i.e. are equal.

We need one more rule called *EXPANSION*. This enables us to generate the equation in Fig.5 from the definition in Fig.7.

$COUNTIMP(n) = [\![MUX | REG(n) | INC]\!] \backslash l1 \ l2$

Fig.7.

In order to state the *EXPANSION* rule we need some notation for abbreviating behaviour expressions. If $X = \{l1, \ldots, ln\}$ is a set of line names then $\{\sim x \sim : x \in X\}$ is an abbreviation for $\{\sim l1 \sim, \ldots, \sim ln \sim\}$. Using this this notation we can state the *EXPANSION* rule as follows:

EXPANSION Suppose $Fi(\underline{xi})$, where $\underline{xi}$ is a (possibly empty) vector of variables, are defined for $1 \leq i \leq m$ by:

$$Fi(\underline{xi}) = \lambda\{x : x \in Xi\}.\{y=Ey : y \in Yi\}, F(\underline{Ei})$$

and $G(\underline{x1},\ldots,\underline{xm})$ is defined by:

$$G(\underline{x1},\ldots,\underline{xm}) = [\![F1(\underline{x1}) | F2(\underline{x2}) | \ldots | Fm(\underline{xm})]\!] \setminus l1 \ldots ln$$

then by EXPANSION we can infer:

$G(\underline{x1},\ldots,\underline{xm}) =$
  $\lambda\{x : x \in (X1 \cup \ldots \cup Xm)-(Y1 \cup \ldots \cup Ym)\}$
  $letrec \{y=Ey : y \in (X1 \cup \ldots \cup Xm) \cap (Y1 \cup \ldots \cup Ym)\}$
  $in \{y=Ey : y \in (Y1 \cup \ldots \cup Ym)-\{l1,\ldots,ln\}\}, G(\underline{E1},\ldots,Em)$

All this rule says is that the behaviour obtained by connecting devices together is got by solving the set of equations resulting from equating the values on input and output lines with the same name. For example, if we take $F1$, $F2$, $F3$ to be MUX, REG, INC as defined in Fig.2, and G to be COUNTIMP defined by Fig.7, then by EXPANSION we immediately deduce the equation in Fig.5 and hence by the manipulations following Fig.5 we conclude that $COUNT(n) = COUNTIMP(n)$ for all n.

Summing up we see that to verify that the device $COUNTIMP(n)$ defined in Fig.7 is a correct implementation of the specification in Fig.4 we just apply EXPANSION (using the definitions in Fig.2) followed by several UNFOLDINGs, followed by REDUCTION, followed by UNIQUENESS.

SPECIALIZATION TO nMOS

We now apply our notation to nMOS circuits. First we restrict all lines to carry just three values: 1 (high), 0 (low) and $\oplus$ (floating). We shall sometimes require, or infer, that lines only carry the two values 1 and 0; such lines will be called boolean. Before defining the behaviour of our nMOS primitives we need to extend the boolean operators $\neg$ and $\vee$ to handle $\oplus$ by defining $\neg \oplus = 1$ and for all $x: \oplus \vee x = x \vee \oplus = x$.

We shall build circuits out of four kinds of primitives: gates (transistors), joins, pullups and ground. Examples of these are shown in Fig.8

Gates have memory: they remember (via their capacitance) the last value input on their control line ($cntl$ in the example in Fig.8). If on the next cycle this line is isolated - i.e. $\oplus$ is put on it - then the stored value controls the gate's behaviour. The output of the gate is $\oplus$ (i.e. 'floats', or is 'isolated') if the controlling value is not high; if it is high then the output value equals the input value (i.e. the gate conducts - is 'on'). This controlling value is the value input on the control line if this is not ⊕, otherwise it is the stored value. Notice that the stored value is updated every cycle, so if a value is being stored and then the control line is isolated, then during the next cycle the stored value will become ⊕. Thus gates have a one cycle decay time - i.e. the stored charge can only be assumed to persist for one cycle. Other decay times are easy to model, see Gordon (1981) for a discussion of this and for more explanation in general.

The other three primitives are all purely combinational. Joins ∨-together the incoming values using the extended ∨ defined above. Notice that joining an isolated line (i.e. one carrying ⊕) to another line does not effect the value on the other line (since $x \vee ⊕ = ⊕ \vee x = x$). Pullups pull ⊕ up to 1. They crudely model the fact that, electrically speaking, if the input to a resistor is isolated then no current flows, hence there is no voltage drop across it and so the voltage at both ends must be equal. Ground gives a constant low.

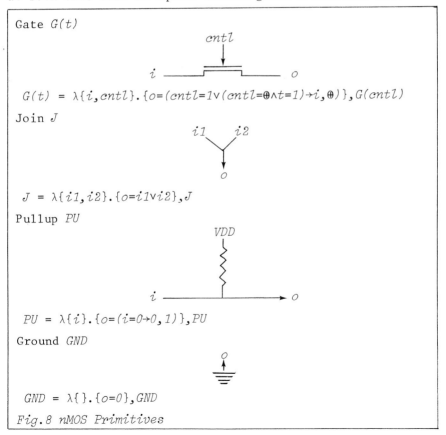

Fig.8 nMOS Primitives

When using the primitives in Fig.8 we will usually need to systematically rename their lines. We could define 'generic' behaviour expressions parameterized on line names, but we will not go into details here. To show that the primitives actually work as expected when connected together we examine examples.

The Inverter   A simple nMOS inverter is:

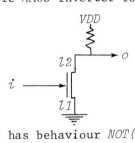

In our model this has behaviour $NOT(t)$ defined by:

$NOT(t) = [\![G(t) | GND | PU]\!] \backslash l1\ l2$

where $G(t) = \lambda\{l1,i\}.\{l2=(i=1 \vee (i=\theta \wedge t=1) \to l1, \theta)\}, G(i)$
$GND = \lambda\{\}.\{l1=0\}, GND$
$PU = \lambda\{l2\}.\{o=(l2=0 \to 0, 1)\}, PU$

Hence by *EXPANSION*:

$NOT(t) = \lambda\{i\}.$
  $letrec\ \{l1=0, l2=(i=1 \vee (i=\theta \wedge t=1) \to l1, \theta)\}$
  $in\ \{o=(l2=0 \to 0, 1)\}, NOT(i)$

and so by *UNFOLDING* and *REDUCTION*:

$NOT(t) = \lambda\{i\}.\{o=((i=1 \vee (i=\theta \wedge t=1) \to 0, \theta)=0 \to 0, 1)\}, NOT(i)$

which by a little 3-valued boolean algebra simplifies to:

$NOT(t) = \lambda\{i\}.\{o=\neg(i=\theta \to t, i)\}, NOT(i)$

Thus non-floating inputs are inverted, and if the input is floating then the negation of the stored value is output. Notice how the use of θ enables us to model the sequential aspects of inverters which are exploited in implementing registers in nMOS. If we require line i to be boolean then the behaviour of $NOT(t)$ simplifies to:

$NOT(t) = \lambda\{i\}.\{o=\neg i\}, NOT(i)$

and this is identical to the purely combinational NOT' where:

$NOT' = \lambda\{i\}.\{o=\neg i\}, NOT'$

since if we define $NOT''(t) = NOT'$ (for all t) then:

$NOT''(t) = \lambda\{i\}.\{o=\neg i\}, NOT''(i)$

and hence by *UNIQUENESS NOT(t)* = *NOT''(t)* = *NOT'*. Thus requiring all lines to be boolean reduces behaviours to the conventional behaviours of classical switching theory.

The NOR gate A simple nMOS implementation of a *NOR* gate is:

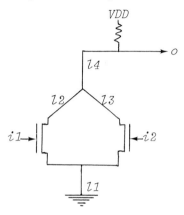

In our model this has behaviour *NOR(t1,t2)* defined by:

NOR(t1,t2) = ⟦*G1(t1)*|*G2(t2)*|*GND*|*J*|*PU*⟧\ *l1 l2 l3 l4*

where

$G1(t1) = \lambda\{l1,i1\}.\{l2=(i1=1 \vee (i1=\oplus \wedge t1=1) \to l1, \oplus)\}, G1(i1)$
$G2(t2) = \lambda\{l1,i2\}.\{l3=(i2=1 \vee (i2=\oplus \wedge t2=1) \to l1, \oplus)\}, G2(i2)$
$J = \lambda\{l2,l3\}.\{l4=l2 \vee l3\}, J$
$PU = \lambda\{l4\}.\{o=(l4=0 \to 0, 1)\}, PU$
$GND = \lambda\{\}.\{l1=0\}, GND$

By *EXPANSION*:

NOR(t1,t2) = $\lambda\{i1,i2\}.$
  letrec {$l1=0$,
    $l2=(i1=1 \vee (i1=\oplus \wedge t1=1) \to l1, \oplus)$,
    $l3=(i2=1 \vee (i2=\oplus \wedge t2=1) \to l1, \oplus)$,
    $l4=l2 \vee l3$}
  in $\{o=(l4=0 \to 0, 1)\}, NOR(i1,i2)$

and so by *UNFOLDING*, *REDUCTION* and some boolean algebra:

NOR(t1,t2) =
  $\lambda\{i1,i2\}.\{o=\neg((i1=\oplus \to t1, i1) \vee (i2=\oplus \to t2, i2))\}, NOR(i1,i2)$

Thus the *NOR* gate outputs the negation of the disjunction of its inputs if these are boolean, otherwise the stored value is used. If we restrict *i1* and *i2* to be boolean then:

NOR(t1,t2) = $\lambda\{i1,i2\}.\{o=\neg(i1 \vee i2)\}, NOR(i1,i2)$

By a similar uniqueness argument to the one used in the *NOT* example it follows that for all $t1$ and $t2$: $NOR(t1,t2) = NOR'$, where *NOR'* is the purely combinational device defined by:

$$NOR' = \lambda\{i1,i2\}.\{o=\neg(i1 \vee i2)\},NOR'$$

Thus showing again that if we ignore ⊕ the model collapses to classical switching theory.

CONCLUSIONS

Alas, we do not have space to give more examples. In Gordon (1981) we prove that the stack cell (Plate 6) and controller (Fig. 3.11) of Mead and Conway (1980) are functionally correct using the properties of *NOT* and *NOR* gates derived above. The proof is highly structured: the controller is expressed as the composition of two subsystems and a clock, and its correctness follows from the correctness of the subsystems. The correctness of these, in turn, follows from the correctness of their immediate constituents, which are the *NOT* and *NOR* gates described above. It is not necessary to flatten down to the 'gate level' and so the proof does not explode in size. However, whether such proofs can be scaled up to realistic examples remains to be seen. It will certainly be an effort to do so, requiring machine assistance. Verified correctness is going to be expensive; will it ever be worthwhile?

ACKNOWLEDGEMENTS

Most of my ideas derive from the pioneering work of George Milne and Robin Milner (see Milner (1980)). Thanks also to Luca Cardelli, Steve Crocker, Igor Hansen, Matthew Hennessy, Jacek Leszczylowski, Ben Moszkowski, Gordon Plotkin and Lee Smith for advice and ideas. This research was supported partly by an S.R.C. Advanced Research Fellowship and partly by The University of Southern California's Information Sciences Institute.

REFERENCES

Gordon, M. (1981). A Model of Register Transfer Systems with Applications to Microcode and VLSI Correctness. Dept. of Computer Science Internal Report CSR-82-81, University of Edinburgh.

Milner, R. (1980). "A Calculus of Communicating Systems". Springer-Verlag Lecture notes in Computer Science.

Mead, C. and Conway, L. (1980). "Introduction to VLSI Systems". Addison-Wesley.

MAGNETIC-BUBBLE VLSI INTEGRATED SYSTEMS

H. Chang*, W. W. Molzen*, J. P. Hwang+, and J. C. Wu+

*IBM T. J. Watson Research Center, Yorktown, NY 10598, U.S.A.

+Carnegie-Mellon University, Pittsburgh, PA, U.S.A.

I. INTRODUCTION

In Mead and Conway's "Introduction to VLSI Systems" [1], it was stated that "... no subsystem can require a different technology for the generation of its internal signals. Thus such fully integrated systems cannot be implemented solely in a technology such as magnetic bubbles, since it cannot create the signals required for all operations in the on-chip medium." This paper disputes the above statement. To rectify this small blemish in the otherwise interesting and incisive book, we shall present bubble logic devices (embodied in a systolic-array string-pattern matching chip), which receive and generate signals in the form of bubbles. In fact, it is also possible to implement all-bubble conservative logic [2], intelligent memory chips, parallel and pipelined processors, PLA's etc. We expect that the magnetic bubble technology will benefit from the Mead and Conway heirarchical design methodology as much as the semi-conductor technologies.

Commercial bubble memory chips [3] already contain one million bits, with on-chip read and write as well as selection, detection, and redundancy functions. Four-million-bit memory chips are expected in 1982, and density advances are expected to remain at a better pace than the semiconductors. Bubbles exceed semiconductors in density and in chip capacity by about 10 times mainly due to the intrinsically simpler device structures (one or two critical lithography levels for memory cells and one additional level for switch and logic gates). Bubble bits are intrinsically in data streams, natural for parallel and pipelined constructs. Moreover, the data flows and gate functions are all synchronized to a common drive (hence self-clocked). The signals in the form of bubbles are replenished in energy and re-normalized during each step of bit movement (i.e. every clock cycle). The abundance of bits, data streaming, self clocking, and signal renormalization appear to be unique advantages of bubble logic.

The major disadvantages are slow speed (10 MHz data rate experimentally demonstrated) and geometrical constraints (step-wise movement of bubbles in a single plane). However, the regular structures to realize parallel and pipe-lined chip architectures such as systolic arrays or programmable logic arrays not only achieve speed improvement but also alleviate geometrical constraints.

II. BUBBLE MEMORY AND LOGIC DEVICES

Bobeck et al. [4] reported perforated-sheet bubble memory devices in 1979, which eliminate bulky coils, thus resulting in compact package and fast speed. We have evolved logic components in this technology. Fig. 1 depicts the operation of a shift register (a storage and propagation element). Concentration of current at the edges of holes creates strong magnetic poles to attract or repel bubbles. By properly phasing the current directions in the two conductor sheets, a train of magnetic poles is induced to propagate bubbles in the magnetic garnet layer underneath.

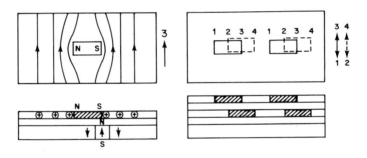

Fig. 1. Perforated-sheet shift register memory cell.

Fig. 2 depicts the structure and function of an AND/OR gate. On top of two parallel propagation tracks (A and B), a third-layer conductor loop is used to activate the logic device (see Figs. 2b and 2g). When both inputs receive bubbles (i.e. A=B=1), Figs. 2a to 2e describe the action sequence of bubble interaction to perform logic. In essence, both bubbles are expanded within the confinement of the conductor loop during phases 4-1. Due to mutual repulsion, the expanded bubbles still remain in their original tracks, thus yielding 1's at outputs A OR B and A AND B. When A=O and B=1, Figs. 2f to 2j describe the action sequence. During phases 4-1, the control current is applied. The track B bubble is prevented from advancing on its own track, but is expanded into track A. The current is turned off at phase 2, when track B repels the bubble while track A attracts the bubble. Thus the bubble contracts into track A; i.e. A OR B=1, and A AND B=0.

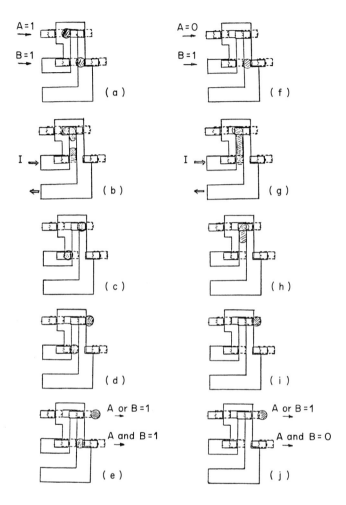

Fig. 2. A perforated-sheet AND/OR gate.

III. STRING-PATTERN MATCHING CHIP

Given a desired pattern $P=p_0p_1p_2...p_k$, and an input text $S=s_0s_1s_2...s_n$, the string-pattern matching operation should yield $r_0r_1r_2...$ where

$$r_i \leftarrow (s_i = p_0) \text{ AND } (s_{i+1} = p_1) ... \text{ AND } (s_{i+k} = p_k), i = 0,1,...,n-k.$$

Foster and Kung [5] have designed an algorithm to execute the above operation. The pattern and the text streams are made to flow against each other so that a bit in each stream can be compared to all bits in the other stream.

Overlapping pairs of bits are compared concurrently. In successive time steps, successive sections of the text stream (K + 1 bits long) are compared to the pattern P.

Only two basic cells are required for the matching chip--an one-bit comparator and an accumulator. See Fig. 3. The comparator transmits the input data streams, (P_{out} ← P_{in}, S_{out} ← S_{in}), but also compares them ($P_{in} = S_{in}$). The comparison result is ANDed with the comparison result of the higher order bit (D_{in}) of the same character.

$$P_{out} \leftarrow P_{in}, \quad S_{out} \leftarrow S_{in}$$

$$D_{out} \leftarrow D_{in} \text{ AND } (P_{in} = S_{in}); \text{ i.e. } D_{in} \text{ AND } (\overline{P_{in} \text{ XOR } S_{in}}).$$

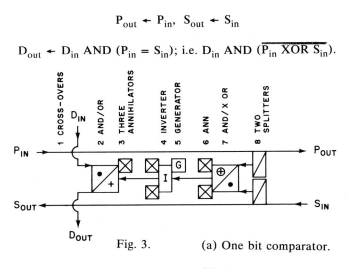

Fig. 3. (a) One bit comparator.

(b) Accumulator.

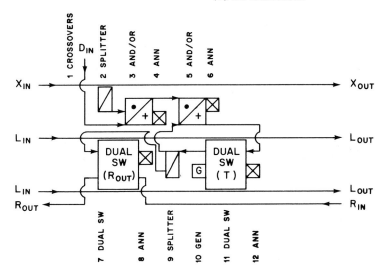

The other basic cell is an accumulator, which receives D_{in} (the result from the comparator above), λ_{in} (the end of pattern indicator), and X_{in} (the don't care bit). It maintains a temporary result T, and at the end of the pattern, uses T to replace the result r that flows from right to left.

$$L_{out} \leftarrow L_{in}; \qquad X_{out} \leftarrow X_{in}$$

If L_{in}, then $R_{out} \leftarrow T$, $\qquad T \leftarrow$ True
$\qquad$ else $R_{out} \leftarrow R_{in}$, $\qquad T \leftarrow T$ AND $(X_{in}$ OR $D_{in})$

While the comparisons $s_i = p_0$, $s_{i+1} = p_1$, $s_{i+2} = p_2$, ... are being executed sequentially, the accumulator performs the AND function and stores the temporary result. At the end of the pattern ($L_{in} = 1$), the final result r_i is released. Note that during the accumulation process, the T_{next} is a function of $T_{present}$. The looping time of the operation $T \leftarrow T$ AND $(X_{in}$ OR $D_{in})$ determines the rate of pipelined processing.

Fig. 4 presents a string-pattern matching chip layout. In the center 64 one-bit comparators are arranged in eight rows (i.e. 8 bits per character) and eight columns ($p_0 p_1 ... p_7 p_0 p_1 ... p_7 ...$). Eight accumulator cells are placed below. Fig. 5 gives an enlarged view of a corner of the chip. The components in the cells are identified by the same letters as used in the flow diagrams of Fig. 3. Only the AND/OR gate was briefly discussed in Sec. II. The other components (as identified in Figs. 3 and 5) are described in Ref. [6].

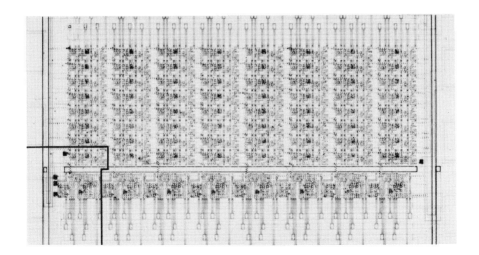

Fig. 4. $\qquad$ Chip layout for string-pattern matching.

The picture shows clearly that data streams (P, S, X, L, R and their duplicates) flow in the horizontal direction (overlapped rectangles in the first two conductor sheets). The logic functions are activated by the vertical conductor loops (third layer). Similar components are aligned vertically to share the same activation conductors.

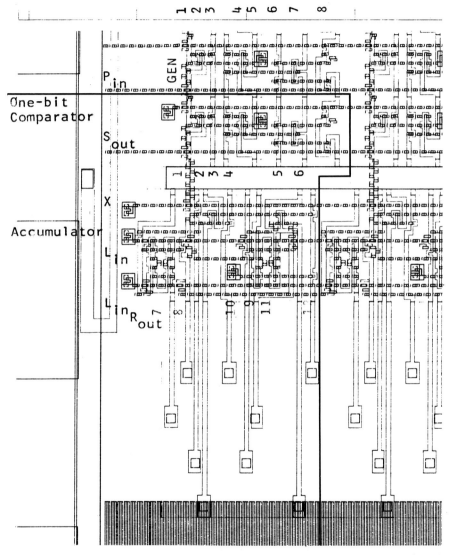

Fig. 5. A corner of Fig. 4 enlarged.

IV. DEVICE DEFINITION AND CHIP LAYOUT

As compared to semiconductor-device chip layout, bubble devices require much more attention to geometrical details. This stems from the basic physical phenomenon of bubble motion. In response to local field gradient, a bubble moves over one spatial period during one time period. Moreover, for fan-out, a bubble must be first stretched by multiples of the spatial period, and then replicated. The communication from one cell to another must assume the contiguity of the cells. These geometrical constraints need not be burdensome to the designers. In fact, with proper formulation at the component definition stage, the chip composition task at a later stage can be greatly facilitated. We have imposed standardization on the various components in terms of dimensions (integer multiples of propagation period) and input/output (both timing phases and locations).

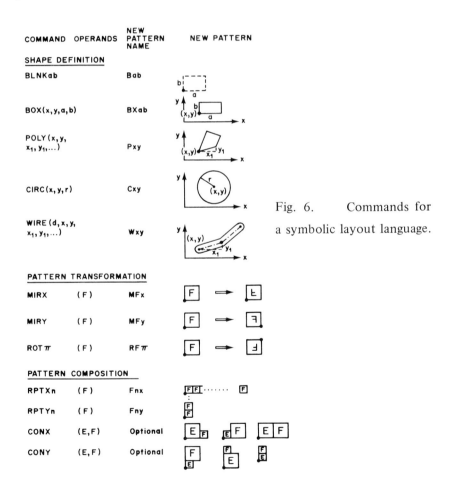

Fig. 6. Commands for a symbolic layout language.

The bubble chip layout task can be viewed as the translation of a data-flow diagram into device geometries (e.g. from Figs. 3 to 5). The data-flow diagram is a functional specification of the chip. Then placement rules can be defined to fit various bubble components along the data flow paths to implement the chip. A symbolic layout language (i.e. a simple set of rules) (1) enables concise and precise documentation, (2) facilitates communication between workers using different CAD systems (for example, CIF [1] based on detailed arithmetical description and GL-1 based on interactive graphics, (3) allows the build-up of modules and hierarchical designs, and (4) expedites verification and analysis after layout.

A small set of commands are given in Fig. 6, which fall into three groups: shape definition, shape transformation and composition. The basic geometrical shapes (with dimensions and locations) include: rectangles (or called boxes), polygons, circles, and wires (i.e. areas covered by moving circles in straight lines).

Table   Symbolic layout language program for a one-bit comparator.

1	RPTX 14	(Px)	--> Pl	26	CONY	(T15, T14)	--> T16
2	BLNK 14 1		--> Bl	27	CONX	(T3, T16)	--> T17
3	CONY	(Pl, Bl)	--> Tl	28	CONX	(MPXy, MPXy)	--> T18
4	CONX	(Cro, Pl)	--> Top	29	BLNK 2 1		--> B6
5	BLNK 1 4		--> B2	30	CONY	(T18, B6)	--> T19
6	BLNK 1 1		--> B3	31	CONY	(T19, T18)	--> T20
7	RPTY 3	(Py)	--> P2	32	CONX	(Axo, T20)	--> T21
8	CONY	(P2, B3)	--> T2	33	CONY	(T21, B4)	--> T22
9	CONX	(B2, T2)	--> T3	34	CONX	(T17, T22)	--> Mid
10	CONX	(An, B3)	--> T4	35	MIRY	(Tl)	--> MTly
11	CONX	(T4, An)	--> T5	36	MIRY	(Cro)	--> MCROy
12	MIRY	(Px)	--> MPXy	37	CONX	(MCROy, MTly)	--> Bot
13	RPTX 3	(MPXy)	--> T6	38	CONY	(Bot, Mid)	--> T23
14	CONY	(T5, T6)	--> TE6	39	CONY	(T23, Top)	--> LEFT
15	CONX	(Aor, TE6)	--> T7	40	CONY	(MPXy, B3)	--> T24
16	BLNK 4 1		--> B4	41	CONX	(Rep, T24)	--> T25
17	CONX	(B4, An)	--> T8	42	CONY	(B6, T25)	--> T26
18	CONY	(T8, T7)	--> T9	43	CONY	(T26, B6)	--> T27
19	CONY	(An, B3)	--> T11	44	CONY	(MPXy, Rep)	--> T28
20	CONX	(Ge, T11)	--> T12	45	CONY	(Tu, Px)	--> T29
21	CONY	(T6, T12)	--> T13	46	CONX	(T28, T29)	--> T30
22	CONX	(Inv, T13)	--> T10	47	CONY	(T27, T30)	--> T31
23	CONX	(T10, T13)	--> T14	48	CONY	(T31, B6)	--> RIGHT
24	BLNK 9 1		--> B5	49	CONX	(LEFT, RIGHT)	--> COMP
25	CONX	(B5, An)	--> T15	50	END		

V. VECTOR-SCAN ELECTRON-BEAM LITHOGRAPHY

As very dense and pseudo-random logic patterns are being pursued for bubble chips, and scaling-down from the present 2μ features to 1μ and 0.5μ is anticipated, vector-scan electron-beam lithography was chosen to generate chip patterns (either on masks or directly on wafers). Our interactive graphics system can feed the pattern design file directly to the EB system, hence we consider the EB design issues at the stage of pattern generation. Electrons are scattered when they hit the resist/substrate composite, and cause shape-to-shape interactions in exposing electron beam resists. For instance, for 0.5μ PMMA EB resist, the back scattering of electrons by the substrate to define one shape may affect shapes several microns away. In order to correct this proximity effect and maintain pattern fidelity one or more techniques (shape sizing, shape partitioning, and dose variation) may be used. For a general introduction to the use of electron beams in microelectronic fabrication, refer to Brewer [7]; and for recent advances in proximity correction, refer to Parik [8] and Molzen and Grobman [9].

Much CPU time is consumed in executing proximity correction algorithms. The following two techniques are used to reduce the CPU time [9]. To maintain well-defined narrow gaps between large shapes the framing technique (only considering the frame of a large pattern in its interaction with other shapes) saves considerable CPU time. To maintain shape fidelity in areas with dense small shapes, the grouping technique (dividing entire area into groups, and using separate dose correction for each group) reduces the CPU time from N^3 dependence to N dependence (N = no. of shapes). At the device design stage, critical gaps between large shapes can be limited, more natural grouping may be provided, and shape insensitive components can be identified.

VI. CONCLUSIONS

1. The hierarchical design methodology (decomposition, abstraction, recursion, etc.) as expounded by Mead and Conway and endorsed by many others has proven highly valuable for the development of semiconductor integrated systems. The careful analysis of a problem and the adoption (or invention) of an algorithm indeed serve as an excellent example of abstraction and decomposition as Foster and Kung have demonstrated. This paper illustrates that the above techniques are equally applicable to magnetic bubbles. In fact, our bubble implementation of the string-pattern matching chip bears close resemblance to Foster and Kung's nMOS chip.

2. The early work of design methodology in semiconductors was restricted to nMOS device structures and fabrication techniques and attention was focused on system designs on a chip. By contrast, we used the system framework to invent a new set of bubble logic components, and evolved standardization and modulation at the component level to adapt to the geometrical constraints of bubbles and to facilitate chip composition later.

3. Bubble logic capability has been known as long as bubble memory capability, but has only been sporadically pursued. We have designed components with identical lithography levels and fabrication steps for both memory and logic, and we have also established the parallelism between bubbles and semiconductors at system chip design level. Thus, we hope to benefit further from the vigorous pursuit of semiconductor VLSI chip designs, and to encourage more rational assessment of the usefulness of intermingling logic and memory in bubble chips.

References

1. Mead C. and Conway L., (1980) "Introduction to VLSI Systems", Addison Wesley, Reading, Mass.

2. Chang H., (1981) "Magnetic-Bubble Conservative Logic", presented at the Physics of Computation Workshop, MIT, May 6-8, 1981; to appear in proceedings.

3. Chester M., (1980) "Magnetic Bubble Memory Update", Computer Design, pp. 232-240.

4. Bobeck A. H., et al., (1979) "Current-Access Magnetic Bubble Circuits", Bell System Tech. Journal, vol. 58, no. 6, pp. 1453-1542.

5. Foster M. J. and H. T. Kung, (1980) "Design of Special-Purpose VLSI Chips: Example and Opinions", Computer, vol. 13, no. 1, pp. 26-40.

6. Chang H., Hwang J. P., and Wu J. C., (1981) "Magnetic-Bubble VLSI Chip Design--Memory and Logic Component Library and Chip Layout", Research Report, Electrical Engineering Department, Carnegie-Mellon University.

7. Brewer G. R. (Editor), (1980) "Electron Beam Technology in Microelectronic Fabrication', Academic Press, N.Y.C.

8. Parik M., (1978, 1979) (a) "SPECTRE, A Self-Consistent Proximity Effect Correction Technique for Resist Exposure", J. Vac. Sci. Technol. 15, 931 (1978); (b) "Corrections to Proximity Effects in Electron Beam Lithography". I. Theory, (c) II. Implementation, (d) III. Experiment", Journal of Applied Physics 50, 4371, 4378, 4383.

9. Molzen W. W. and Grobman W. D., (1981) "A Model for Optimizing CPU Time for Proximity Correction", to be published in J. Vac. Sci. Technology.

SESSION 3

THE INMOS HARDWARE DESCRIPTION LANGUAGE
AND INTERACTIVE SIMULATOR

Brian Collins and Alan Gray

*Inmos Ltd., Whitefriars, Lewins Mead
Bristol BS1 2NP*

1. INTRODUCTION

This paper describes a hardware description language and simulation system, currently in use at Inmos, which are intended to support the development of very large MOS circuits. It also gives some of the considerations influencing their design.

The language (called simply HDL) allows designers to specify the structure and connectivity of a VLSI design concisely and accurately. Although the language is purely declarative and does not specify action or procedures in any way, its design has been strongly influenced by modern software high-level languages such as Pascal and Ada.

The simulator combines the ability to simulate circuit designs at circuit (analogue), logic (timing) or behavioural levels, or at any combination of levels, from the same circuit description (in HDL). Any module in a design can, in principle, be simulated at any level, with the simulator automatically providing thresholding and idealising at the interfaces between the various levels of simulation. For example, a critical circuit may be simulated at circuit level within a logic simulation of its environment to provide a realistic, but efficient, test-bed.

The user interface of the simulator is highly interactive and is strongly modelled on an interactive software debugger, allowing (or even encouraging) the designer to 'debug' his hardware. The designer can interrupt, resume and restart simulations, can inspect results prior to producing prints or plots, and can interactively change the values of signals. The designer may set up circuit or logic waveforms as input and may request output in many formats to the terminal, a

printer or to a file. Traps on the values of logic signals
may be established to set up 'breakpoints' for important or
erroneous conditions.

2. HARDWARE DESCRIPTION LANGUAGE

Requirements

To support comparatively large teams designing VLSI
circuits with more than 100,000 devices, the following
requirements were identified.
- That descriptions in the language be both human and
  machine readable.
- That the language be a good means of communicating
  designs between co-operating designers.
- That the language be powerful enough to highlight major
  design features but still allow the minutest detail (eg.
  transistor or interconnect dimensions) to be specified.
- That the language encourage structured, regular and
  modular design techniques.
- That the language be easy to read.
- That the language be easy to write.
- That common hardware constructions be supported in the
  language by concise notation.
- That the language support the features of the simulator
  and other, as yet unspecified, utilities.

Language features

Almost all of the commercially available simulators have
an input format that bears more resemblance to a symbolic
assembler than to a high-level language. They do not seriously attempt to meet any of the requirements above.

By contrast, the hardware description language HDL is a
modular high-level language designed to satisfy all of the
requirements.

The language features:-
- A hierarchical module structure. Modules are the fundamental elements of HDL and are designed (defined) by
  interconnecting instances of smaller modules. Modules
  are analogous to procedures or functions in software
  languages.
- Signals which may be given names and are global or local
  to a module. A signal is a signal name, the output of a
  module (function), a concatentation of signals, or a
  signal repeated a number of times. Signals may be
  explicitly connected together giving names to the results
  of modules or aliases to other signal names.

- Parameters which name the connections from a module to the outside world. Module definitions have formal parameters which are signal names and are analogous to the formal parameters of software procedures. Module instances have actual parameters which are (arbitrarily complex) signals and are analogous to actual parameters in software procedure calls, The compiler checks that the number and size (number of bits) of the actual parameters match those of the formal parameters specified in the module definition.
- Attributes of modules, instances, parameters or signals which are 'keyword=value' pairs. These attributes augment the topological structure represented by the modules and signals by giving physical or logical properties (eg. transistor dimensions or logic delays) to the item to which they refer. Attributes may be specified on user-defined modules, possibly with default values, with the values specified individually on each instance.
- A library of predefined modules which are in-built primitive elements from which all modules are ultimately composed. The primitive elements are those known to the simulator and other utilities. The library includes:
  - Passive circuit elements (resistors, capacitors).
  - Transistors of various type (N & P channel, enhancement & depletion).
  - Logic gates (NAND, NOR, INV)
  - Matrix gates which specify a structure of similar transistors that provide a single (but complex) logical function. HDL has a special notation to handle the actual parameters of matrix gates.
- Behavioural specifications which allow the operation of a module to be described in some detail for logic simulation before a silicon implementation has been designed, or for verification of a proposed implementation.

Example

Fig. 1. gives an example which aims to give some feel for the structure of HDL without going into too many details. It defines two modules, REG which is a single four bit register, and REG_FILE which includes sixteen four bit registers and an address decoder. The modules NE and UNE are N-channel enhancement transistors, predefined in the library. INV is an inverter, and DECODE4 is a four to sixteen bit decoder assumed to be defined elsewhere. The example makes strong use of a shorthand which allows instances of the library modules to be passed parameters of any size causing automatic replication of the instance across the whole width of the parameters.

```
MODULE REG (IN  input [0:3], clock_, enable,
            OUT output[0:3])

    SIGNAL gated_clock = UNE (enable, clock_)

    output = UNE (INV (gated_clock@[0:3]), input),
             UNE (gated_clock@[0:3], INV (output))

END REG
```

```
MODULE REG_FILE (IN  addr[0:3], input[0:3],
                     clock_, write_enable,
                 OUT output[0:3])

    SIGNAL control[0:15] = DECODE4 (addr),
           ram[0:15][0:3]

    FOR i = [0:15] DO
        NE  (control[i]@[0:3], ram[i], output),
        REG (input, clock_,
             UNE (write_enable, control[i]), ram[i])

END REG_FILE
```

Fig. 1. A short example of HDL

Multiple Signals

Apart from its high-level modular structure, HDL differs from other hardware languages by its very powerful treatment of multiple signals (buses) and repetition of signals, instances and connections. Signals (including parameters) may be given one or more ranges (one per dimension) which are not necessarily contiguous. Eg.
   SIGNAL x[0:7], y[15:0] , z[0:3, 8:11]
These signal names may subsequently be used without an explicit range to indicate the whole bus, or with a range to indicate some, not necessarily contiguous,part of it. E.g.
   x[0:3]   -- the first nibble of x
   x[7:0]   -- the whole of x in reverse
   x[0:7:2, 7:0:2] -- all the even bits of x followed by
                  -- all the odd bits of x in reverse.
Similarly the ranges used in a repeated statement (the

FOR clause in Fig. 1.) can be as general.

3. THE INTERACTIVE SIMULATOR

To support INMOS' designs of VLSI MOS circuits the following requirements were identified for the simulator or simulators that would be needed.
- That the simulator be able to simulate at any of the following levels:
  - Behavioural, in which blocks of circuit are described in terms of their logical operation and timing without implying a specific implementation in silicon.
  - Logical, in which the designed hardware at the transistor or logic gate level is simulated using designer specified timing for propagation delays.
  - Circuit, in which the individual transistors and diffusion diodes are simulated at an analogue level using theoretical or semi-empirical models for charge flow and capacitances.
- That the simulator use the same high-level input language to describe a circuit for simulations at all levels.
- That the designer be able to simulate different parts of the circuit at any of the above three levels in parallel without specifying additional pseudo-circuitry to perform voltage thresholding or idealisation between levels.
- That the designer be able to interactively inspect and modify a simulation allowing him to 'debug' his simulation.

No commercially available simulator meets these requirements so at Inmos we have designed and implemented our own.

The input language for all levels of simulation is HDL. This allows the designer to interconnect instances of the primitive modules known to the simulator. Behavioural descriptions in HDL can be simulated behaviourally or can be used to verify the correct operation of an implementation. The same library of primitive modules can be simulated both at logic and circuit levels. Therefore logic simulations model the bidirectional properties of pass transistors and resistors, and in circuit simulations logic gates are 'macro-expanded' to their constituent transistors. This macro-expansion restricts the available logic gates to those that have a single obvious implementation in terms of transistors for the selected technology, ruling out AND, OR and XOR gates.

Logic Simulation

Logic simulation is event driven; transitions on a signal which is an input to a primitive module cause the outputs of the module to be re-evaluated after a propagation delay, which in turn may cause subsequent transitions on other signals. There is no fixed time-step at which events happen; the simulation is purely free-running.

Signals simulated at logic level have one of three levels: high (H), low (L), and unknown (U), the latter representing an intermediate level. Signals also have one of three strengths: forcing (F), non-forcing (N), and high-impedance (Z). Therefore the logic value of a signal is one of the nine products of a level and a strength.

Signals can be simultaneously driven by more than one module, requiring the competing values to be resolved. When driving values are of different strengths, the stronger value predominates. When two or more of the strongest values are of different levels, the resultant level is unknown (U), otherwise the value of the signal is that of the strongest driving signal.

Signals which are not being driven by any modules retain their level with high-impedance strength. Such high-impedance signals may decay after a time to unknown (U) level as the charge 'leaks' away.

Similar rules are used to determine the resultant value of two signals connected together by a switched-on transistor. For a bidirectional transistor, extra rules ensure that the transistor is only driving in one direction at a time. For a transistor stated to be 'unidirectional', checks are made that it is not 'driven backwards', ie. when switched on, its input must always be at a stronger value than its output.

This accurate logic modelling of primitive devices together with the more conventional logic gates makes the logic simulator a powerful but relatively inexpensive tool for the design of static or dynamic MOS circuits.

Circuit Simulation

The circuit level algorithm is conventional, using a modified Gauss-Seidel iteration to solve the circuit matrix at each time-point, and a semi-heuristic adaptive time-step.

The system permits the coexistence of multiple models for non-linear devices, any one of which may be used at the designer's discretion in a particular simulation. Currently, the MOS transistor model used has a DC characteristic based on Foss et al. (1979) and uses the capacitance model of Meyer (1971).

The junction diodes modelling source and drain active areas are lumped together where a single node acts as the source and/or drain of several transistors, so that the total count of non-linear elements in the circuit is kept to a minimum.

The designer is able to interactively vary the model parameters and examine the changes in device characteristics which result. Specific sets of parameter values may be dynamically created, modified and deleted with two levels of mnemonic naming.

The first of these, referred to as the subtype name, is used as an attribute in the HDL text, either explicitly or by default, eg. to distinguish enhancement from depletion MOS devices. A subtype name is mandatory for all device instances in the HDL and for all sets of parameter values.

The optional second level name, referred to as the box name, may be used to select slightly different parameter values for the same general type of device depending on its dimensions, eg. the width and length of a MOSFET gate. Although names are given in the simulator to sets of parameter values distinguished in this way, no reference need be made to box names in the HDL text as the proper selection is performed automatically from the device dimensions given.

Multi-level Simulation

There are two main problems that arise in implementing multi-level simulations. These are:
- maintaining synchronisation of the logic and circuit parts with respect to external input waveforms, to scheduled output prints and plots, and to each other.
- correctly resolving the logic value or voltage on a signal that is at the interface between logic and circuit simulations and may be driven by either part.

Synchronisation is achieved by making the circuit simulation part a sub-component of the logic simulator. Between logic events(which include scheduled inputs or outputs) the circuit simulator is permitted to run until the time of the next logic event. The circuit simulator either runs to completion for this time or stops at some earlier time if a signal at the logic/circuit interface crosses some threshold.

The circuit simulator preserves its state across events and therefore can maintain its own time-step which is independent of any times in the logic simulator (eg. plot periods). This ensures that the synchronisation overhead between the simulators is negligible.

Signal Resolution

The interface between circuit level and logic level simulations models the driving/load impedance of the elements on the logic side by means of a simple RC pair, in which the capacitance is computed statically from the number and type of logic elements connected to the interface and the resistance is selected dynamically from the strength of the logic signal as one of the three values (see Figure 2 below). The level of the logic signal is modelled by a voltage source which takes one of three values. Since this form of logic/circuit interface requires nothing to be added to the HDL, the user may choose to simulate the same hardware description at either circuit or logic levels, or any combination of the two, and may place the dividing lines between logic and circuit simulation at any point that is suitable for the degree of accuracy or speed of simulation he wishes to achieve.

Voltage changes in the circuit simulation which cause the interface node to cross the logic threshold trigger level-changing events in the logic simulation. Only forcing high and low values are generated by circuit outputs, since the circuit simulation is assumed to be the more accurate.

Separate threshold voltages may be specified for low-to-high and for high-to-low transitions. Some positive separation of these thresholds prevents high-frequency spurious transitions during periods when the interface node is metastable. The resistance values used in modelling the three logic strengths may also be adjusted interactively.

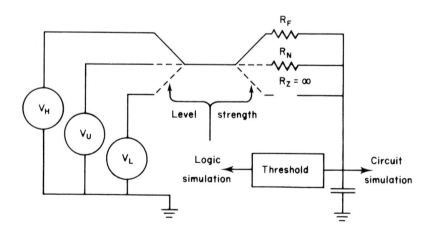

Fig. 2. Logic/circuit interface.

Interactive Control

To reduce the effort required to control the simulation and to make the simulator more pleasant for the designer to use, an interactive screen-based system is used. The screen is split into two parts. The top part is a three line scrolling screen editor in which commands to the simulator are constructed. The editor is an in-house standard editor so the same user interface is preserved across a number of programs. The command on the current line is obeyed when carriage return is pressed. Previously entered commands may be edited and re-obeyed. The rest of the screen is an independently scrolling 'glass teletype' in which all commands are echoed and all terminal output appears. The contents of this lower screen are preserved and may be printed.

A running simulation may be interrupted from the terminal. Any commands may be given at this stage, allowing the designer to inspect or modify signal values, to create or cancel input waveforms or outputs, to resume the simulation or to restart with different initial conditions, or just to give up.

Commonly used commands or command sequences can be allocated to individual function keys so that they may be invoked with a single key press.

4. CONCLUSIONS

From our experience in designing, implementing and using the system certain features stand out as having particular significance.

A high-level language, with good repetition facilities, which allows the connectivity to be statically checked by a compiler, greatly increases the reliability, readability and ease of writing of circuit descriptions for simulations.

Multi-level simulations are easy to specify and control since no extra circuitry is required to perform any interfacing. They significantly reduce simulation times as parts of the circuit are only simulated to the level of detail required.

Interactive control of the simulator also increases the speed at which simulations are performed. The ability to edit commands and re-obey them is particularly useful.

Overall, we have produced a language and simulator that meets our requirements better than any commercially available alternative. This system is now in active use supporting the development of Inmos' new VLSI designs.

5. REFERENCES

Foss, R.C, Harland, R. and Roberts, J., AN MOS transistor model for a micro-mini computer based circuit analysis system, *European Solid State Circuits Conference 1979.* pp 56-57

Meyer, J.E. (1971). MOS models and circuit simulation. *RCA Rev.*, Vol 32, pp 42-63

A Pragmatic Approach to Topological Symbolic IC Design

Neil Weste
Bryan Ackland

Bell Laboratories
Holmdel, New Jersey 07733

1. INTRODUCTION

Symbolic layout systems which use a fixed grid and a character based symbol set have been in use for some time now [Gibson and Nance 1976; Larson 1978; Clary *et al.*, 1980]. Symbolic layout systems which use a notion of free-form topological placement and mask synthesis (commonly called a "sticks" approach [Williams 1978]) are currently uncommon. These latter systems offer advantages over fixed-grid systems with respect to total freedom from geometric design rules, improved graphical interaction capabilities, packing density and capture of designer intent. In particular there are few existing symbolic design systems which treat the task as any more than a physical design problem, with symbols being used to reduce the complexity of geometric design rules rather than totally alleviating them.

This paper examines a proven design system (given the name *MULGA*) for the design-rule free symbolic layout and verification of MOS integrated circuits. Special attention is given in this paper to the *chip assembly* phase of the design process. As a result of experience gained through designing a number of integrated circuits, solid design methodologies have evolved. These are also described along with a summary of circuits designed to date.

2. SYSTEM OVERVIEW

The *MULGA* system, as illustrated in Figure 1, is primarily used to support *structured-layout* [Mead and Conway 1980] based on custom or semi-custom cells. A major advantage of the system is the simplicity and ease of human interaction brought about by the use of simple languages to describe the various stages of design and a high-performance raster-scan color display. Design is supported in the physical, structural and behavioral domains. This allows designers who are neither software nor layout specialists to complete designs with a high degree of confidence.

The chip design methodology developed for use in the *MULGA* system is based on the fact that custom cells may be easily constructed and verified. This approach is in contrast to the idea of a standard cell system in which predefined physical, structural and behavioral models exist for a limited set of SSI functions. To date very little in the way of a library of cells has evolved. Rather, a library of templates representing functions, which a designer uses to mold into a particular custom cell has resulted.

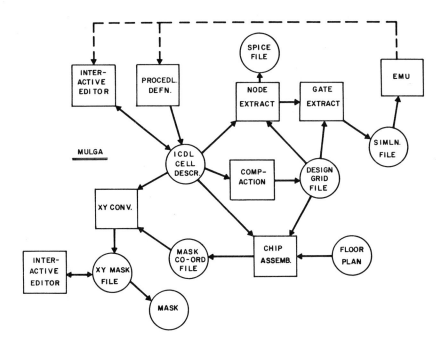

Figure 1. Mulga Software Modules

The chip design methodology can be divided into six phases:

- Floor plan generation
- Functional Subdivision
- Cell Synthesis
- Physical Layout Generation
- Simulation
- Optimization

The first step, generating a floor plan, involves identifying the major functional blocks and then arranging these to minimize the amount of communication between blocks. In addition, global decisions about power supply, control and data routing are made. For instance, a convention of running control vertically in polysilicon and data horizontally in metal (mostly) might be adopted.

The second step is recursive and is responsible for breaking the major functional blocks into manageable cells that may be custom designed. Where possible, use is made of hierarchy to define common sub-blocks for higher level customization and repetition for array structures. Once the lower level cells have been defined one may synthesize these cells. Section 3 will cover those methods used in the *MULGA* system.

The generation of a physical layout to follow a floor plan (commonly known as *chip assembly* [Mudge *et al.*, 1980]) is where this particular chip design methodology parts company with existing methods. This step is responsible for transforming the symbolic description of cells to a valid set of masks. It is complicated by the fact that cells interact with each other. Some of the methods for managing this are covered in Section 4.

Simulation and functional verification are necessary both at the cell and the chip level. This is also described in Sections 3 and 4.

Optimization takes advantage of the fact that cells are easily specified and functionally verified. One may find, for example, that a cell has an aspect ratio that does not blend with its neighbors. The solution is to rearrange the cell so that its aspect ratio changes. This is really where the human mind does excel and the interactive potential of the *MULGA* system is realized. This leads to an interesting philosophical point which relates to "automatic" or "artificially intelligent" systems. In order to bound such a system a set of ground rules need to be defined. When these ground rules can not be satisfied, the systems usually resort to human input. The human response to an obstacle of this type is to redefine the ground rules. An interactive approach allows the designer to use initiative in solving problems which transcend many levels of abstraction. This is why an interactive approach is favored at this time as a practical means of dealing with the broadest range of IC design problems.

This overall chip design methodology is similar in principle to that which has been used by designers of large custom chips for quite some time. The difference here is that all design may be completed in the symbolic domain. The main advantages of dealing with the circuit in this manner are:

- In common with cell design, a geometric design rule free description is generated. Thus the process independence of a cell may be elevated to the chip level.

- The amount of data to be manipulated is kept to a minimum. One symbolic source file generates all subsequent data.

- Increased use of hierarchy. One parent symbolic cell may generate many sibling mask level variants.

3. CELL DESIGN

Once a floor plan has been specified, the process of cell design can begin. Cells form the basic building blocks of the integrated circuit. They represent not only the electrical primitives used to implement circuit function, but also the topological primitives used to satisfy overall circuit connectivity. Some cells lend themselves to a procedural definition - examples include memory arrays, address decoders, routing cells etc. Random logic cells are more easily designed using interactive techniques. Then again a few special cells are conveniently designed from a simple textual description.

In a unified system, all of these constraints and design styles are merged by the cell design language. Geometric languages explicitly define physical topology but ignore electrical function. High-level procedural languages, on the other hand, accurately characterize circuit function at the expense of connectivity and placement. In the *MULGA* system, a symbolic language known as ICDL (Intermediate Circuit Description Language) [Weste 1981a] is used in all phases of cell design. ICDL defines a cell as a distribution of electrical elements over a *virtual grid* as shown in the example of Figure 2. Primitive elements of the language include *devices, wires,* and *contacts.* Note that these elements have physical and functional properties, unlike the inert rectangles and polygons of a geometric language. Two additional primitives, *cell instances* and *pins,* complete the set. *Cell instances* enable the designer to hierarchically build large cells out of smaller ones. *Pins* are named connection points which link a cell to other cells in the circuit. They add structural intent to the design and play an important role in placement, routing and verification.

The primitive elements of a cell are linked both electrically and topologically through the concept of the *virtual grid.* This relative placement network defines layout topology without specifying the physical distance between elements. An implicit electrical connec-

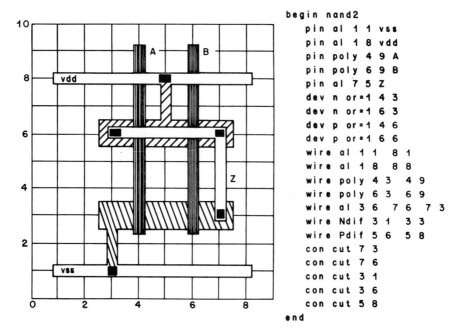

Figure 2. Text and Graphic ICDL Cell Description

tion is made between two primitives existing on the same layer at the same virtual grid position. This is based on Buchanan's *coordinode* concept [Buchanan 1980]. Wires thus form an implicit single layer electrical connection between virtual grid points. Contacts, on the other hand, connect several layers at one grid point. Devices connect to three adjacent grid points as shown in Figure 3.

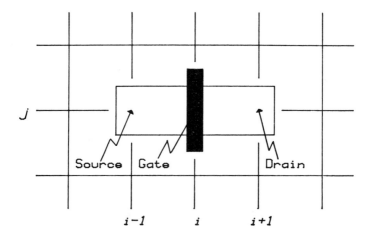

Figure 3. Transistor Virtual Grid Connection Points

ICDL is a simple "assembly-level" language in which each statement corresponds to exactly one circuit element. Procedural definitions and parameterization are not supported at this level. This greatly simplifies the design of software tools such as interactive editors and circuit extraction programs (The complete system is hosted by a DEC LSI 11/23). It also allows for direct textual interpretation by the designer. The simplicity of ICDL, however, does not limit its usefulness in high-level structured design approaches. Rather, it defines a "common ground" meeting point where all cell design aid programs come together.

3.1 Synthesis

As mentioned previously, cells can be designed interactively, procedurally or textually. The *MULGA* interactive graphics editor supports interactive cell design on a high performance color display station [Weste and Ackland 1980]. Features of the display station include extremely fast response and the ability to pan, in real time, around a large symbolic data base.

Procedural definition is accomplished using the *C* programming language to directly generate ICDL files. Experience has shown, however, that the number of procedurally defined cells has steadily decreased as the interactive tools have improved. An experienced user can typically design a new cell interactively in less than an hour. Procedural definition is therefore mostly reserved for large repetitive structures such as PLAs and decoders.

3.2 Verification

In contrast to the many I.C. design tasks which have been automated, cell design still benefits from the creative interaction of a human designer and is thus susceptible to human error. Hence the need for rigorous functional verification at this level. Verification is generally a two-stage process consisting firstly of automatic circuit extraction and secondly of functional simulation. Symbolic design techniques simplify the extraction task by introducing structural or circuit information in place of unnecessary geometric data. In particular, ICDL carries implicit circuit connectivity data - independent of geometric interpretation.

In *MULGA*, two programs perform circuit extraction. The first identifies unique electrical nodes within the circuit and generates a net list description of the cell. It also calculates parasitic capacitance values for each node and combines these with specified device parameters to produce a transistor level circuit description suitable as input to *SPICE* [Nagel and Pederson 1973]. The second program performs gate decomposition, converting series/parallel combinations of transistors into scaled combinatorial gates. This leads to a higher-level circuit description which can be interpreted by a timing simulator.

Analog circuit simulators such as *SPICE* give accurate reliable feedback and the expense of large amounts of computer and engineer time. They typically run on large time-share facilities which are physically distant from the design station. In an interactive design environment, ease of operation and fast turnaround are preferred to absolute accuracy in order that the designer may be encouraged to use the facilities. Accordingly, a simple MOS timing simulator known as *EMU* has been included as part of the resident *MULGA* software [Ackland and Weste 1981].

Timing simulators model digital circuits as collections of idealized transistors which may be grouped in a predefined manner to form simple logic functions. Unlike logic simulators they model analog variables and are able to deal with limited analog effects such as charge storage and bidirectional circuit elements. Performance, however, is typically one to two orders of magnitude faster than their analog counterparts. Experience has shown that timing simulation is accurate to within process parameter variations for all but the most demanding situations. Even dynamic memory cells, whose character is essentially analog, are correctly modeled.

3.3 Compaction

Cells designed using a free form symbolic language such as ICDL contain functional and topological descriptions but no explicit size or spacing information. Not only does this simplify the design process but it also creates a "design-rule free" data-base. To convert a symbolic design into a physical mask description, it is first necessary to assign absolute physical coordinate values to symbolic grid positions. This is conveniently done in two steps. First the cell is shrunk to its minimum physical size as determined by process design rules. This is known as *compaction*. Second, the compacted cell is expanded to satisfy global interconnection constraints as specified by the chip floor plan. This is part of *chip assembly* and is described in Section 4.

A number of compaction strategies have been proposed in the literature [Akers et al., 1970; Dunlop 1978,1980; Hseuh and Pederson 1979]. Fixed grid compaction is a process which assigns a fixed minimum spacing between adjacent grid lines in the symbolic layout, removing those grids (or portions of grids) that are vacant. Non-fixed grid algorithms allow objects to break away from the grid structure and move freely within limits imposed by the assigned topology. Non-fixed algorithms generally give much better compaction at the expense of huge amounts of computation time.

MULGA uses a technique of *virtual grid* compaction [Weste 81b] in which elements are constrained to their original grid positions but the grid is allowed to contract (or expand) in a non-uniform manner. Each grid line is spaced from its parallel neighbors according to the elements that have been placed on those grids. Compaction begins by first examining spacings in one direction only (say the X direction). Each X grid line is compared with neighboring X grid lines for interference. An interference is said to occur if two elements on two neighboring X grids have the same Y coordinate value. The spacing needed to avoid this interference is determined using process design rules. Overall X grid spacing is then defined to be the greatest interference spacing found between the grids. Note that oblique interference is ignored during this first phase.

During the second phase, Y grid spacings are determined in much the same way. This time, however, oblique interferences are included using the X spacings calculated in the first phase. A special ICDL file is generated to highlight critical compaction points within the cell. Using the interactive editor, the designer is able to see those element pairs whose interference was responsible for setting coordinate spacing. The designer may then go back and alter the cell topology to alleviate those critical points. Experience here has shown that this feedback is most useful when a designer is first using the technology. With a little practice, it is possible to design cells that are near optimum the first time. In some situations of course, for example in the design of a RAM cell, every last micron counts. This feedback process is then used even by the experienced designer to achieve the best possible compaction result.

4. Chip Assembly

As was explained in Section 3, an ICDL cell is compacted using the concept of a *virtual grid*. If a cell is to be used in isolation then the design grid file resulting from the compaction may be used to generate the minimum size physical embodiment of that cell. With the chip design methodology used in this system, however, cells are usually aggregated to form larger cells until the major functional blocks comprising the chip are defined. This means that abutting cells which have common interconnection points in the structural domain must be locally warped to maintain communication in the physical domain. This is called *pitch matching* and, to date, three methods have evolved for completing this task:

- Global Pitch Matching
- Local Pitch Matching
- Weighted Local Pitch Matching

Global pitch matching was the first method evolved and is also the simplest. Basically, one constructs orthogonal slices of cells for the major functional blocks and compacts these as single entities. For example, to globally pitch match cells A,B and C in Figure 4, a super-cell ABC is constructed. This cell is then compacted, with the Y design grid file of ABC being used to expand cells A,B and C in the Y direction. The disadvantage of this method is that a worst case expansion tends to result unless all (vertical) design grid tolerances are compatible for all cells (A,B and C).

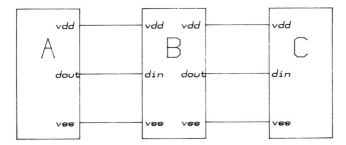

Cells to be Pitchmatched

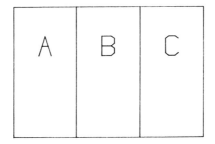

Composite Cell

Figure 4. Pitch Matching of Cells

Localized pitch matching recognizes that abutting cells need only match physically at designated connection points. Non-connecting grids are therefore free to move about locally to optimize local cell spacing. In the example of cells A,B and C shown in Figure 4, only the symbolic connection points illustrated need be matched. Thus the procedure adopted is to individually compact each cell. A matching program then examines the designated connection points between cells and adjusts the design grid files so that the abutting cells will match.

Weighted localized pitch matching is a refinement will be introduced to automatically deal with the interface between a densely packed cell and a relatively sparse cell. A typical example is that of a control cell abutting the end of a register array. If both cells are compacted separately one will compress to a much smaller size than the other. If a matching point is designated, it is possible for the smaller cell to have abnormally oversized elements and to actually increase the overall size of the composite cell. To correct this, sparse cells will use additional information from the compacter to reduce the length of critical components (such as diffusion wires) and ignore artificial spaces caused by sparse cells.

When two cells are joined by abutment it is necessary to ensure that both structural and physical design rules are obeyed. This is achieved by either instituting rules defining how cells may be joined, or by compacting a cell in its "environs". The first method is easier to implement, although the latter is the more correct. Two rules of abutment that have been used to date are:

- If a cell abuts itself (as in an array of similar cells) then elements on the line of abutment should be symmetrical.
- Any points that match topologically in two dissimilar adjoining cells must be declared connection points.

4.1 Language

In describing the chip assembly language the following terminology is used. A *p-cell* is a parent cell defined in ICDL. An *s-cell* is one particular instance of a parent cell and is termed a sibling cell. In addition to inheriting the parent name, an s-cell has a name to distinguish it from all other siblings that have the same parent. To the chip assembly software, an s-cell is a rectangular box with an ordered set of named *pins* that may exist on each of the four sides of the periphery of the cell. Pin naming relates the s-cell name, the ICDL pin and the connection face. A typical pin name is as follows:

s-cellname.ICDLpinname.direction

where

— *s-cellname* is the sibling cell name

— *ICDLpinname* is the parent cell pin name in the ICDL description

— *direction* is one of N,S,E,W to unambiguously specify a connection face
This form of a pin name will be called *p-name* in the following descriptions.

In our first attempt at a *chip assembly* language, the following statements and associated operations are allowed:

— **PLACE** *p-cellname x y s-cellname*
Place p-cell called *p-cellname* at symbolic *x,y* and call it *s-cellname*.

— **NORTH** *s-cellname1 p-cellname s-cellname2*
North of sibling *s-cellname1* place a cell whose parent is *p-cellname* and call it *s-cellname2*. Similarly for **SOUTH, EAST** and **WEST**.

— **LABEL** *p-name newname*
Place a new pin in the ICDL description of the assembled cell called *newname* at all the instances of *p-name*.

— **CONNECT** *p-name1 p-name2*
Try to connect pin *p-name1* to *p-name2*. This is tried in a variety of ways. Firstly the pins are tested to determine whether they pitch match. If they do not, the cells are slid to allow matching. Failure here results in an error message.
Random routing channels are generated symbolically as special ICDL cells which are simply included as part of the chip floor plan. To aid the designer specify busses

and in general reduce the amount of data that has to be input, standard pattern matching characters are used.

4.2 Example

To illustrate this language a simple example will be given. The following file TEST (describing the cells shown in Figure 5a) is input to the chip assembler.

```
PLACE A 0 0 a
EAST a B b
EAST b C c
CONNECT a.vss.E b.vss.W
CONNECT a.vdd.E b.vdd.W
CONNECT a.dout b.din
CONNECT b.vss.E c.vss.W
CONNECT b.vdd.E c.vdd.W
CONNECT b.dout c.din
LABEL a.din DIN
LABEL c.dout DOUT
```

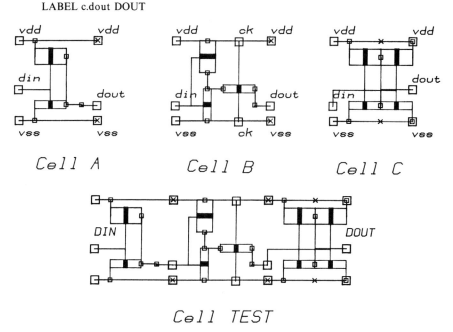

Figure 5. Component Cells before and after Chip Assembly

The following design grid match file is generated:

drmatch 0 Y   A B C   (i.e. match cells A,B and C at symbolic Y=0)
drmatch 1 Y   A B C
drmatch 5 Y   A B C

In addition, the following ICDL file is generated (shown in Figure 5b):

```
begin TEST
   instance A 0 0 "a"
   instance B 5 0 "b"
   instance C 11 0 "c"
   pin poly DIN 0 2
   pin poly DOUT 16 2
end
```

A check plot indicating the connections made is also generated, as shown in Figure 6.

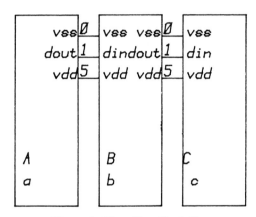

Figure 6. Floor Plan Check Plot

5. EXPERIENCE

To date five chip designs have been completed using the cell design and chip design methodology described in this paper. These are summarized in the table below.

Chip Designs Completed				
Name	Tech.	No.Devices	Errors	Design Time(man-weeks)
RALU	CMOS	5000	1	8
ADC	NMOS	1000	0	3
DIFF	NMOS	1000	0	3
DTW	CMOS	14000	-	24*
XCL	CMOS	1800	-	6*

* (currently being fabricated)

The one error that has occurred so far was a missing contact in the first CMOS design. This was caused by the oversight of a reported error in the connectivity check due to the desire to commission the first circuit. All designs met timing and functional design specifications (including the RALU chip which fortunately only suffered the loss of one function due to the missing contact).

This one error did teach a very valuable lesson. No matter how well the design aids monitor performance, all is for naught if this information is ignored. As a result of the experience gained on the first design, procedures were instigated for automatically updating the various data representations when the original ICDL file is edited or modified.

5.1 Designer Expertise

A reduction in the complexity of the IC design process is important for two reasons:
- designers with no previous experience are encouraged to think in terms of LSI (or VLSI) solutions to systems problems
- experienced designers may take a custom design approach to complex problems

The following points are representative of the expertise needed to design CMOS circuits on the *MULGA* system:
- It is assumed that the designer knows what circuit is needed for a particular function. If not, the designer is given a set of canonic CMOS SSI functions in circuit form and perhaps some prototypical ICDL layouts.
- A layout style is suggested to the designer. For CMOS we encourage designers to use the Gate-Matrix style [Lopez and Law 1980] with deviations where necessary. This style uses vertical polysilicon gate signals and horizontal transistors. Metal runs vertically and horizontally to complete connections.
- A designer is introduced to ICDL with the following circuit layout rules:

    — No direct connection of N-thinox and P-thinox (diffusion)

    — No polysilicon to thinox connections

    — No contacts over gates

    — No implicit transistors (i.e. poly over thinox)

    These rules are supported by the circuit extractor and in the future will be incorporated into the ICDL editor.
- The designer is then taught about substrate contacts.

At the end of this initial education a designer is able to enter cell layouts, compact them, extract circuits and, with minimum effort, simulate a design. After mastering cell layout, the designer may progress to larger cell designs and full chip layouts. This does, of course, require slightly more schooling and experience. In line with this, we have found the simulator to be an excellent teaching tool (in addition to its use as a verification tool). Capacitance on critical lines may be varied interactively and the drive capability of gates changed to optimize performance. This may then be incorporated into the original ICDL description.

5.2 Implementation Stages

Based on our experience with *MULGA*, we can suggest a sequence for system development that will result in a constantly improving design capability from the outset of implementation.

- First define a language suitable for the target technologies. The language must contain elements for which the designer can define a physical, structural and behavioral models. Initially make the language simple, but with some notion of how it might be expanded.
- Write a parser and define an internal data structure for the language. Develop simple data-base management routines.
- Construct general routines to plot and print elements of the language. Build programs using these to display designs. Designs may be constructed procedurally or via a conventional text editor.

- Develop a compaction program and a mask conversion program. A short cut here is to use the system in a *fixed-grid* mode, thus delaying the need for a compacter. Mask descriptions may then be generated.
- Write a circuit extractor and appropriate filters to convert the output for circuit analysis or simulation programs that might be available. If none are available write your own.
- Using the print and plot software write a simple editor which can insert, delete and modify elements in cells. This may be improved incrementally by adding commands and functions.
- Develop a chip assembler

6. CONCLUSIONS

We have described a practical symbolic approach to cell and chip design. Key features of the system include simple unified description languages, a friendly interactive software environment and a self-contained design station based on a high performance color display. An important driving force in the development of the system has been the concurrent design of various integrated circuits. The result is a significant reduction in the complexity of the design of custom I.C.s, this being of particular value to non-specialist designers.

REFERENCES

Ackland, B. and Weste, N.(1981) Functional Verification in an Interactive Symbolic Design Environment, *Proceedings of the 2nd. Caltech Conference on VLSI,* Pasadena CA, 1981.

Akers, S.B., Geyer, J.M. and Roberts, D.L.(1970) IC Mask Layout with a Single Conductor Layer, *Proceedings 7th. Design Automation Workshop,* San Francisco, 1970, pp. 7-16.

Buchanan, I.(1980) *In* Modelling and Verification in Structured Integrated Circuit Design, PhD Thesis, University of Edinburgh, 1980.

Clary, D., Kirk, R. and Sapiro, S.(1980) SIDS- A Symbolic Interactive Design System, *Proceedings of the 17th. Design Automation Conference,* June 1980, pp. 292-295.

Dunlop, A.(1978) SLIP: Symbolic Layout of Integrated Circuits with Compaction, *Computer Aided Design,* Vol. 10, No. 6, Nov. 1978, pp. 387-391.

Dunlop, A.(1980) SLIM- The Translation of Symbolic Layouts into Mask Data, *Proceedings of the 17th. Design Automation Conference,* June 1980, pp. 595-602.

Gibson, D. and Nance, S.(1976) SLIC- Symbolic Layout of Integrated Circuits, *Proceedings of the 13th Design Automation Conference,* June 1976, pp. 434-440.

Hseuh, M.Y. and Pederson, D.O.(1979) Computer-Aided Layout of LSI Circuit Building Blocks, *Proceedings of the 1979 International Symposium on Circuits and Systems,* July 1979, pp. 474-477.

Larson, R.P.(1978) Versatile Mask Generation Techniques For Customer Microelectronic Devices, *Proceedings of the 15th Design Automation Conference,* June 1978, pp.193-198.

Lopez, A.D. and Law, H.F.(1980) A Dense Gate Matrix Layout Style for MOS LSI, *IEEE Journal of Solid State Circuits,* Vol. SC-15, No. 4, Aug. 1980, pp. 736-740.

Mead, C. and Conway, L.(1980) *in* Introduction to VLSI Systems, Addison-Wesley, Reading MA.

Mudge, J.C., Peters, C. and Tarolli, G.M.(1980) A VLSI Chip Assembler *In* Design Methodologies for Very Large Scale Integrated Circuits, NATO Advanced Summer Institute, Lovain-la-Neuve, Belgium, July 1980.

Nagel, L.W. and Pederson, D.O.(1973) Simulation Program with Integrated Circuit Emphasis, *Proceedings of the Sixteenth Midwest Symposium on Circuit Theory*, Waterloo, Canada, April 1973.

Weste, N. and Ackland, B.(1980) An IC Design Station needs a High Performance Color Display, *Proceedings 17th. Design Automation Conference*, Minneapolis MN, June 1980, pp. 285-291.

Weste, N.(1981a) MULGA- An Interactive Symbolic Layout System for the Design of Integrated Circuits, *Bell System Technical Journal*, Vol. 60, No. 6, July-Aug. 1981, pp. 823-857.

Weste, N.(1981b) Virtual Grid Symbolic Layout, *Proceedings of the 18th. Design Automation Conference*, Nashville TN, June 1981.

Williams, J.(1978) STICKS- A Graphical Compiler for High Level LSI Design, *Proceedings of the NCC*, May 1978, pp. 289-295.

A Bit-Serial VLSI Architectural Methodology for Signal Processing

Richard F. Lyon

VLSI System Design Area, Xerox Palo Alto Research Center, 3333 Coyote Hill Road, Palo Alto, CA 94304 U.S.A.

INTRODUCTION

Applications of signal processing abound in the modern world of electronics. Telephones, stereos, radios, and televisions are the most common examples. In the general electronic communication and control market, there is a widespread demand for higher quality and lower cost signal processing components of all sorts. For many years, digital signal processing (DSP) techniques have been touted as "the way of the future," but have consistently failed to make a big impact on the market. We discuss here an *architectural methodology* designed to make digital signal processing "the way of the present."

An architectural methodology is a style, or school of design, that provides a basis for a wide range of architectures for different functions. The architectural methodology presented here is built on top of the logic, circuit, timing, and layout levels of VLSI system design methodology presented by Mead and Conway (1980). It includes a large component that is independent of the underlying technology.

Researchers working on DSP theory and applications have had a hard time implementing their ideas in cost-effective hardware, because the IC industry has not figured out how to support their needs. We hope to change that, based on the family of components described, in conjunction with the ability for designers to easily create their own more specialized system components from the silicon layout macros and composition rules that characterize our methodology. The standard-chip way of life has caused many inappropriate architectures to be proposed and tried in the past; the new freedom for non-specialists to easily design custom integrated systems will enable a wave of new applications and new architectures.

SIGNALS and SIGNAL PROCESSING

Signals are time-varying measurements or simulations of real-world phenomena; for example, an audio signal represents minute changes in air pressure as a function of time. In most familiar signal processing equipment, a signal is represented by an electrical analog, such as a continuously changing voltage; such analog signal representations can be directly processed through continuous-time components such as resistors, capacitors, inductors, diodes, amplifiers, etc. The modern alternative is to sample the signal at equally spaced instants of time, and to represent those sampled values either as discrete-time electrical analogs (e.g. amount of charge held on a capacitor) or as numbers (in some kind of a digital computing machine).

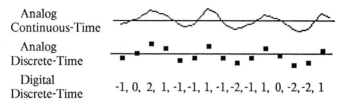

Analog Continuous-Time

Analog Discrete-Time

Digital Discrete-Time   -1, 0, 2, 1, -1, -1, 1, -1, -2, -1, 1, 0, -2, -2, 1

The theory of sampling and discrete-time signal processing is quite well developed, and applies to either analog or digital representations of signals. Switches, capacitors, and amplifiers are typical building blocks for analog discrete-time signal processing. For digital signal representations, the basic building blocks are memories, adders, and multipliers. Analog noise, analog drift, and digital roundoff effects of these components are also well understood.

Digital representations other than sequences of sample values are sometimes used; for example, delta-modulation and its variants use a fast sequence of one-bit values indicating whether the signal is above or below its predicted value at each time instant. These representations will not be considered here.

Using modern MOS VLSI technologies, large amounts of signal processing can be done with a small amount of silicon, compared to the more mature continuous-time analog technologies. Yet, signal processing devices remain relatively expensive and in various ways limited. We believe that difficulties in both design and usage of these devices is the reason. Therefore, we base our architectural decisions on our desire to jointly minimize design difficulty and usage (programming) difficulty, subject to maintaining the high performance promised by the technology. We also believe that it is not possible, for the range of applications we are considering, to get good enough performance from analog VLSI technologies; nor is it easy to design the required analog circuits. Accordingly, we have arrived at these basic decisions:

Question: *Answer:*

Continuous-time or discrete-time?   Discrete
    (so that MOS VLSI techniques, particularly digital techniques, can be used)

Analog or digital sample representation?   Digital
    (to achieve highest quality and tractable design)

Bit-parallel or bit-serial number representation?   Bit-serial
    (for high clock rate, and maximum flexibility and extensibility)

Bus-oriented or dedicated signal paths?   Dedicated
    (for maximum extensibility, and efficiency)

General-purpose programmability or specialized?   Specialized
    (for efficiency and ease of application)

Fixed or variable filter parameters?   Variable
    (to cover the widest range of applications, including time-varying filters)

What number system?   Fixed point 2's comp. LSB first
    (for efficiency and ease of logic design)

DIGITAL FILTERS

Probably the most commonly needed signal processing component is a filter. It is simply a linear (usually time-invariant) system with memory (i.e. the output is a linear function of some history of the input). Most commonly, the input and output are single scalar signals, though this is not required; complex- and vector-valued signals will not be discussed, but are easily accommodated in this architecture. In a digital discrete-time architecture, a filter is a computation that operates on a sequence of numbers as input, and produces another sequence of numbers as output. The usual purpose of such computations is to pass some frequencies of signals, while attenuating others. See Moore (1978) for an introduction to the mathematics of digital filters and related signal processing components; for more detailed information, including four chapters on hardware implementations, see Rabiner and Gold (1975).

COMPONENTS AND HIERARCHY

The remainder of this paper explains how to implement signal processing components (operators and systems) in a style evolved from that presented by Jackson, Kaiser, and McDonald (JK&M) in their 1968 paper "An Approach to the Implementation of Digital Filters." JK&M's notion of an *approach* is similar in some respects to our more developed notion of an architectural methodology, but lacks the notions of standardized interfaces and hierarchical composition of operators.

Our system-building strategy is to design top-down, decomposing blocks functionally into more detailed block diagrams, until we get down to low-level operators, such as adders and multipliers; then, to construct the system from the bottom up, by assembling operators into higher-level operators. To do this, we need conventions for the hierarchical definition and construction of operators. This methodology does not severely constrain the range of possible architectures, but unifies many architectures, allowing them to share components at many levels. The general features of the methodology and conventions are described in following sections.

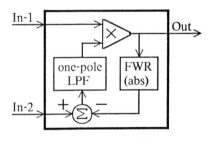

An automatic gain control operator, composed of other operators.

The architectural methodology involves the use of heavily pipelined bit-serial arithmetic processors, all operating at a fixed throughput rate (words per second), and *multiplexed* over several channels or functions to match processing hardware rate to the signal sample rates. In this approach, it is important that the throughput of an adder be the same as that of a multiplier, which is in turn the same as that of a second-order filter section, or a delay element, or any other component (and the same as that of a wire, the elementary data path).

MULTIPLEXING

Digital time-division multiplexing is a technique developed by the Bell System to send several digital sampled data signals on a single wire pair (one-bit data path), by dividing time into periodically recurring time-slots, each of which could carry one signal sample. The number of signals that can be carried on one data path is then the word rate divided by the signal sample rate (the word rate is the bit rate divided by the number of bits per word, since a data path carries one bit at a time at a fixed rate). Bell's T1 carrier system is a good introductory example; it carries 24 voice-band signals, sampled at 8000 samples per second (8 ksps), with 8 bits per sample, on a wire pair with 1.544 Mbps total data rate (that's 24*8000*8 plus a few extra bits). These multiplexing concepts are easily extended from transmission of signals to processing of signals.

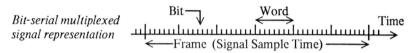

Bit-serial multiplexed signal representation

For many operators, multiplexing is as simple as time-interleaving the samples of several input signals, resulting in interleaved answers. But when the definition of an operator requires some *state* associated with a signal, multiplexing that operator requires that more state be saved, separately for each of the interleaved input signals. In digital signal processing, all state information is conventionally saved inside the unit-delay operator (called Z^{-1}, the inverse of the unit advance operator of Z-transform notation), which simply produces an output value equal to what the input value was one sample time earlier. The problem of saving state reduces to the problem of multiplexing the Z^{-1} operator. This operator must store one word for each multiplexed signal, so it is logically just a long shift register (if the signals being multiplexed do not all have the same sampling rate, things are more complicated, but still tractable).

CONVENTIONS

System-wide clocking, signalling, timing, and format conventions are needed in order to hierarchically design a system of potentially very high complexity. The basic approach we have adopted is to have system-wide synchronous clocks at the bit rate, and both centralized and distributed timing and control signal generation, as described below.

Clocking

The synchronous clocking scheme will include several versions of the bit clock, provided to accommodate the different types of technology from which the system will be constructed. These technologies and clock types are (1) LSTTL with positive edge-triggered clocking, and (2) high-performance NMOS with nonoverlapping two phase clocking. The relative clock phasing and data timing across technology boundaries will be fixed system-wide. In the NMOS part, the first clock gate encountered by a signal entering a subsystem will be designated Phi-2 (Phase-In), which will be high during the latter (low) part of the TTL clock cycle, when data bits are stable. Phi-1 (Phase-Out) will be the last clock gate encountered by a signal leaving an NMOS subsystem, and it will be timed such that output data, stable during Phi-2, meets the TTL setup and hold requirements (see chapter 7 of Mead and Conway 1980).

Signalling

Signalling refers to the electrical representation of bits on wires. We adopted the electrical convention that a low voltage level would represent numerical digit 1, and a high level would represent numerical digit 0 (so that unconnected pulled-up inputs default to a zero signal). This is true within and between operators of any scale (subsystems, chips, or whatever). Bits are transmitted with a non-return-to-zero (NRZ) representation, which just means that the voltage changes to the voltage representing the new bit value after the appropriate clock edge, and remains throughout the clock period.

Timing signals are all active-high; the higher voltage state means logically true, asserting the named time state (e.g. the *LSBtime* signal discussed below is high during the least significant bit time of a word).

More electrical conventions for between-chip signalling are embodied in the I/O pad designs used in the NMOS parts. An informal statement of the conventions is that the NMOS parts be "compatible" with LSTTL (fanout of 10) in levels, noise tolerance, and transition times. The signalling conventions are about the only conventions that would need to be changed to accommodate new and different VLSI technologies as they become available. Some changes in clocking conventions may also be desirable at some point.

Timing

The centralized control signal generator is a time counter that keeps track of bits, words, etc. on a cycle equal to the slowest period of interest in the system. This control section may be thought of as a microprogram store with no conditional branching, a very wide horizontal control word output, and a long linear program (but the actual PLA encoding will be much more efficient).

The distributed control scheme is more widely used, and is more suitable for actually addressing the large-system issues, where design of a single controller for the many functions would be a task beyond the capabilities of a human designer, and would add too many interrelated constraints between parts of the system.

The basic distributed control strategy is that each data path be associated with a timing signal (which is used and generated locally) that represents time within a word; optionally, data paths may be associated also with other timing signals that represent time within a larger *frame*, as appropriate for a particular operator. The standard representation of time is simply for the signal to be true during the first time-slot of its frame, and false at all other times. The signal called *LSBtime* is true during the first (least significant) bit of each word, and the signal *Word0time* is true during the first word of each frame, for example.

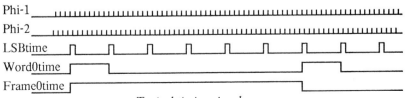

Typical timing signals

Most arithmetic processing units need only clock, data, and one control input to identify the first bit of a new word. Thus, we adopt a convention to use an *LSBtime* signal in association with *every* operator input and output: operators accept *LSBtime* as a control input, and produce a delayed version as the *LSBtime* of their output(s). Thus data paths from one unit to another carry their own timing information, and units can be arbitrarily interconnected without concern about connection to a central controller. Still, the designer must take care to assure that merging signal paths (e.g. at an adder) and loops (e.g. in a recursive filter) have the correct total delay (instead, a special stretchable queue could be designed to automatically adjust delays, but this would greatly decrease the efficiency of the methodology at the chip level).

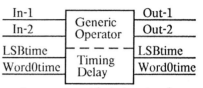

An operator with timing signals.

Format and numerical value restrictions

Some operators, such as multipliers, may take two input operands of different lengths. In such cases, the convention is that both operands are aligned to the same *LSBtime* signal (i.e. words of different lengths are right-adjusted). For short words, the bit slots between the sign bit of one word and the LSB of the next should be filled with sign extensions. Then values of words of any length can be described by the two's-complement binary integer interpretation of the entire 24-bit word (and can be written in binary, octal, or decimal as convenient).

Operators usually require that input values be limited to a range which is less than the total number of bits would allow. For example, the multiplier design for 24-bit words requires that the X input be in the range $[-2^{21}, 2^{21})$, while the Y input must be in the range $[-2^{2k-1}, 2^{2k-1})$, where k is the number of actual hardware stages in the multiplier (not over 12); both X and Y conform to the same format (24 bits per word, LSB first, two's complement), but have different value restrictions. Value restrictions can be enforced by either scalers or limiters, if necessary.

On consistent formats for signals and coefficients

We have deliberately chosen to have signals and coefficients conform to the same format, with their distinction being only in possibly different value restrictions. Many signal processing architectures do not have this property, thereby making it difficult to multiply (mix) two signals together, or to use a filter's output as another filter's coefficient input. These uses of multipliers are not the common time-invariant linear system uses, but are nevertheless widely needed for nonlinear and time-varying processing such as modulation, demodulation, automatic gain control, adaptive filtering, correlation detection, etc. Of particular importance to us is the application of a time-varying filter for speech synthesis; the coefficients of the synthesis filter are themselves the outputs of lowpass (interpolation) filters.

On interpretation of coefficients as fractional values

We have spoken of signal samples with integer values, but we often need to consider the representation of fractional values, especially for coefficients. The interpretation of operands as fractional values derives from the definition of the function performed by the multiplier. For example, the "5-level recoded" or "modified Booth's algorithm" pipelined multiplier design of Lyon (1976) computes $XY/2^{2k-2}$ for a k-section layout, with Y restricted to $[-2^{2k-1}, 2^{2k-1})$. So if we simply regard Y as a value in $[-2, 2)$, we can say that the multiplier computes XY (as long as we interpret X and XY consistently, either both as integers or both as fractions). Another way of saying this is that a multiplier which accepts n-bit coefficients regards the radix point as being $n-2$ places from the right. Multipliers can easily be designed with other scale factors, to give other range interpretations to the operands. The scaling described here is popular for signal processing.

Pipelining delay and composition of operators

Another important convention is that all operators have some constant positive integer number of clock cycles of delay (pipeline delay or transport delay) between the LSB of an input word and the LSB of the corresponding output word. For some operators, such as adders, each output bit depends only on the corresponding and lower-order input bits, so the delay can be very small—but not zero. We require that no output bit be combinationally related to any input bit, so at least one cycle of delay is required. This allows the system clock rate to be fast, independent of how many adders and other operators are cascaded (long ripple propagation delays are excluded, unlike typical bit-parallel systems).

The delay of an operator is required to be a constant, known at design time, rather than being dependent on control parameters. The designer must satisfy delay constraints, such as equal *LSBtime* for signals that merge in an adder, by inserting null operators with carefully chosen delay wherever needed (i.e. shift registers are added to make signals line up in time). Loop delay constraints are sometimes difficult or impossible to satisfy if operators in the loop have too much delay. Fortunately, all delay information is known at design time, and is easy to represent and design with, by using simple notations.

FUNCTIONAL PARALLELISM and MULTIPLEXING ALTERNATIVES

We have briefly discussed the use of multiplexing to accommodate many separate operations on relatively slow signals by using a small amount of fast hardware. In the other direction, *functional parallelism* can be used to perform operations on very fast signals, by using a larger number of hardware operators running in parallel, with appropriate techniques for combining their partial computations. Thus, it is possible to apply the serial-arithmetic architectural methodology to very wideband applications, such as radar and video processing.

There is really no limit on the bandwidth obtainable, as illustrated by the VFFT series of Fourier transform processors described by Powell and Irwin (1978), which use slow 3 Mbps (custom PMOS) serial chips to compute transforms of up to 10 Msps radio telescope data (i.e. the sample rate is actually faster than the bit rate!).

The techniques for use of functional parallelism in nonrecursive (finite impulse response) filters are straightforward. Similar techniques for recursive filters have been demonstrated by Moyer (1976). The point is that the bit-serial architectural techniques do not overconstrain the system-level architecture, and do not limit the applications to low-bandwidth areas.

Six performance regimes

We are familiar with signal processing system designs that span six distinct regimes, in terms of the amount of multiplexing and/or functional parallelism that they incorporate. Several chips mentioned below have been designed by members of the author's team at Xerox, and will be described in forthcoming papers.

The low-performance regime (level 0) is characterized by a lack of multiplier hardware or other special signal processing features, and is exemplified by software systems running on simple general-purpose processors.

Bit-serial architectures begin to look interesting at level 1, characterized by the use of a single multiplier for different functions. This level is exemplified by Jim Cherry's "Synth" chip, a tenth-order lattice filter for speech synthesis; its block diagram has two multipliers per lattice stage and one overall gain control multiplier, while its implementation consists of one multiplier, one adder, one delay shift register, a little control logic, and a handful of data path switches.

When more total performance is required, level 2 is appropriate. This is basically the simple multiplexed filter approach described by JK&M, which uses one hardware multiplier for each multiplier in the block diagram of a section, but shares them over several sections by multiplexing. Lyon's 32-channel second-order "Filters" chip exemplifies this regime; it has almost no data path switching or control logic. Rich Pasco's "FOS" (first-order section) is another example, which uses scalars and switches instead of multipliers, for a range of specialized applications.

For signals of higher bandwidth, the multiplexing factor may be reduced to unity. We call this level 3, meaning one hardware filter per signal, with no sharing. Rich Pasco's "NCO" chip is a rather simplified example; it is a first-order integrator with no multiplexing, so the signal sample rate can be very high (500 ksps).

When the application requires sample rates higher than the word rate obtainable with the target technology, more parallelism is needed. At level 4, systems take several bits per clock cycle into a distributor, which de-interleaves and sends simple serial data to several operators in parallel; their results can later be combined into the net answer, in a partial-result combiner. Jim Cherry's "FIR" chip is intended to be used this way in the implementation of two-dimensional video filters for image understanding. The chip is one of the operators to be run in parallel; the associated distributor, control, combiner, and line memories are quite simple.

Finally, when the sample rate is higher than the achievable bit rate (level 5), another level of parallelism is needed, as in GE's "VFFT-10" system mentioned above. In this "very fast Fourier transform" system, blocks of signals are transmitted along many serial paths in parallel; signal samples are delivered on a collection of data paths by some faster signal source that can deal with the high bandwidths.

THE BUILDING BLOCKS — LOGIC DESCRIPTION

When designing the logic of many of the low-level and system-level operators, it is not important to know what technology will be used to implement those operators. Thus, this level of the design is relatively long-lived and independent of scale of integration, etc.

For most operators, such as adders, multipliers, and filters, we design logic in the style popularized by synchronous edge-triggered TTL MSI families; that is, the exact nature of the clocking is suppressed, and all memory is in D-type flip-flops, sometimes connected as parallel registers or shift registers. Some of the operators, however, such as the larger memories, cannot be adequately represented by a logic-level design; most of the design work for those parts is in the technology-dependent circuit and layout level.

Given logic designs of this form, it is straightforward to apply a collection of techniques to translate them into circuits for the target technology. The techniques we use for NMOS are those described by Mead and Conway (1980), making heavy use of pass-transistors and dynamic storage to produce designs that are much more efficient than gate-based circuits. The logic designs therefore do not imply any particular gate structure or circuit design. Examples of CMOS implementations of multipliers and serial memories for digital filtering can be found in Ohwada, et. al. (1979).

There is not room here to cover details of logic designs, so we just mention some the more useful operators that have been built. They are signal combiners (adder, multiplier, and variations and combinations), point operators (scaler, overflow detector/corrector or limiter, full-wave and half-wave rectifiers, square, and sine), serial memories (Z^{-1} block, random-access serial register file, and serial ROM), and filters (configurable biquadratic sections, parallel FIR blocks, lattice structures, and multipurpose first-order sections for zero-order hold, interpolation, differentiation, sum-and-dump, oscillators, etc.)

An interesting aspect of the memory-intensive operators is the serial-parallel-serial (SPS) memory organization for both read-write and read-only memories, random or sequential access, that simultaneously minimizes area and power while keeping a high input and output bandwidth. Since memory is always accessed for full words, the parallel internals of the memory can operate on a cycle that is much slower than the serial bit clock, allowing minimum-size dynamic memory cells and slow, low power addressing and bus logic. Only the input and output shift registers and a small control PLA need to run at the bit rate.

ON VLSI LAYOUT STYLES

When actually creating layouts for a system designed through our architectural methodology, it is important to use a simple layout style that facilitates placement and interconnection of components, at many levels of the hierarchy. For NMOS, we have chosen a standard grid style with fixed width (200λ) and variable height cells. A component may occupy any number of 200-lambda wide cells, subject to chip size constraints. The grid may be thought of as a distribution network for VDD, Ground, and two clock phases, into which components and wires are dropped (possibly automatically).

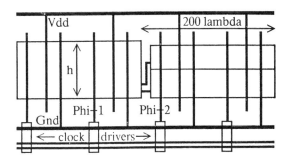

This style is comparable to the polycell layout style, except that the cells being placed and interconnected are considerably larger than gates and registers, and are already specialized to our architectural methodology. This allows considerably more efficient layouts than are possible with polycells, by leaving room for flexible optimizations of logic, circuits, and layout of low-level operators. System designers need never deal with this intra-component level of optimization, but there is always the opportunity for "wizards" to design efficient new low-level components to add to the available library.

Some operators are not designed on the 200-lambda grid, for sake of their layout efficiency. For example, the serial sequential-access memory used in the "Filters" chip is designed with a custom cell array style, appropriate for the large collection of memory cells and the controller PLA.

CONCLUDING REMARKS

As JK&M predicted in 1968, serial arithmetic is alive and well as a simple and efficient way to implement digital filters and other signal processing systems in silicon. By extending their approach into an architectural methodology, and by combining this with a VLSI design and implementation methodology, we suddenly enable designers everywhere to build their own novel low-cost signal processing systems. The methodologies provide a sensible context for the accumulation of a library of component and system designs, at logic and layout levels. So start now, and design the future accordingly.

REFERENCES

Jackson, L. B., Kaiser, J. F., and McDonald, H. S. (1968), An Approach to the Implementation of Digital Filters, *IEEE Trans. on Audio and Electroacoustics* AU-16, 413-421 (*reprinted in* Rabiner and Rader, *Digital Signal Processing*, IEEE Press).

Lyon, R. F. (1976), Two's Complement Pipeline Multipliers, *IEEE Trans. on Communications* COM-24, 418-425 (*reprinted in* Salazar, *Digital Signal Computers and Processors*, IEEE Press).

Mead, C. A. and Conway, L. A. (1980), *Introduction to VLSI Systems*. Addison-Wesley, Reading, Massachusetts.

Moore, F. R. (1978), An Introduction to the Mathematics of Digital Signal Processing, *Computer Music Journal* Vol. 2, No. 1 and 2.

Moyer, A. L. (1976), An Efficient Parallel Algorithm for IIR Filters, *IEEE International Conference of Acoustics, Speech, and Signal Processing*.

Ohwada, N., Kimura, T., and Doken, M. (1979), LSI's for Digital Signal Processing, *IEEE Journal of Solid-state Circuits* SC-14, 214-220.

Powell, N. R. and Erwin, J. M. (1978), Signal Processing with Bit-serial Word-parallel Architectures, *Real-Time Signal Processing*, Proc. SPIE Vol. 154, 98-104.

Rabiner, L. R., and Gold, B. (1975), *Theory and Application of Digital Signal Processing*. Prentice-Hall, Englewood Cliffs, New Jersey.

A NETWORK FOR THE DETECTION OF WORDS
IN CONTINUOUS SPEECH (*)

Jean-Pierre BANATRE, Patrice FRISON, Patrice QUINTON
IRISA, Campus de Beaulieu, Av. du General Leclerc,
35042 RENNES-CEDEX FRANCE

1. INTRODUCTION

These last few years, much work has been devoted to the problem of continuous speech recognition [1]. This research has led to techniques allowing very good recognition rates (>95%) for languages with a limited vocabulary (some hundred words). However systems using these techniques are still much too slow in order to be widely developed.

Recent developments in microelectronics (VLSI circuits [2]) lead to a cheap implementation of highly performant algorithms. For these reason, we think that algorithms should be reconsidered in order to give them a highly parallel structure which may be adapted to new technologies.

The present study is concerned with a word-spotting algorithm that allows to compare the words of a given vocabulary against the outputs of the phonemic analysis of a pronounced sentence. The word-spotting algorithm that we consider has been defined by BAHL and JELINEK ([5]) and a simplified version has been implemented in the Keal Speech Understanding System for the recognition of large vocabularies ([7],[4]). This version constitutes the basis of our work.

The parallel execution scheme of this algorithm combines classical multiprocessing and pipelining. The computation is decomposed into a great number of elementary activities that may be run in parallel. An important characteristic of these activities is that they are all built from the same unique model. In the following, we show that this computation may be supported by a simple and regular processor array which is capable of pipelining

---

(*) This work is supported by the CNET contract nr. 7935226.

searchs of the successive words of the vocabulary.

Section 2 gives a formal description of the reference algorithm. The structure of the basic processor capable of running the elementary activity is proposed in section 3, where it is also shown how such processors have to be connected in order to perform the search of a set of words into the pronounced sentence. Finally, section 4 gives a brief review and discussion.

2. A WORD-SPOTTING ALGORITHM

2.1 The problem

Speech recognition or understanding generally comprises several steps. During a first step, acoustic features are extracted from the digitalized speech waveform. A second step called phonemic analysis is then performed, during which elementary sounds (called abusively phonemes in the following) are determined and identified. Phonetic analysis translates the pronounced sentence into a sequence of N phonetic segments. Let us denote by $(X(1), \ldots, X(N))$ this sequence. Each segment $X(j)$ is itself constituted of a set of phonetic labels which represent the most probable phonemes for this segment. To each phonetic label of a segment is attached a probability. In a third step, the words of the task language vocabulary are matched against the phonetic translation of the uttered sentence. In all generality, every word must be matched against every sub-sequence of the utterrance; this process, referred to as word-spotting, is very time-consuming and needs be controlled by the use of other linguistic knowledges or by heuristics that may reduce the whole search.

Word-spotting consists in searching through the sentence (coded as explained above) all the occurences of a given word represented by the sequence of its constituent phonemes. In other terms, let $Y = (y(1), \ldots, y(M))$ be the sequence of phonemes representing this word. The problem consists in matching Y with every subsequence $(X(k+1), \ldots, X(k+P))$ extracted from X and in evaluating the likelihood of this matching.

While computing this matching, possible errors from the phonemic analyzer have to be considered. Phonemic analysis of a given phoneme z may result in three different cases, namely, confusion, deletion and insertion. A confusion appears when the phoneme is translated into a segment made of several alternative phonemes in which z may or may not be present. An insertion occurs when the phonemic analyzer produces superflous segments. Finally a phoneme may be omitted when the phonemic analyzer does not detect a given phoneme and consequently does not produce the corresponding segment.

2.2 The algorithm

The principle of the word-spotting algorithm is based on a probabilistic model of the behaviour of the phonemic analyzer (PA in the following).

Assume that the word $Y=y(1)\ldots y(M)$ is to be matched against the sentence $X=(X(1), \ldots, X(P))$. Assume that the last phoneme $y(M)$ of Y is a special symbol] called end-of-word marker. To the word Y is associated a Probabilistic Finite State Machine (PFSM in the following) which modelizes the behaviour of PA when dealing with Y. Figure 1 depicts the PFSM associated to a two-phonemes word $(y(1),y(2),])$. This PFSM has four states $S(0)$, $S(1)$, $S(2)$ and $S(3)$. State $S(i)$ ($0 \leq i \leq 3$) describes the behaviour of PA when dealing with phoneme $y(i+1)$. PA may either:
- remain in state $S(i)$ with some probability $Pi(y(i+1))$ and produce some phoneme x with conditional probability $qi(x/y(i+1))$: this modelizes the insertion case;
- enter state $S(i+1)$ with probability $Pc(y(i+1))$ and produce some phoneme x with conditional probability $qc(x/y(i+1))$: this is the confusion case;
- enter state $S(i+1)$ without producing any output, with probability $Po(y(i+1))$: this is the omission case.

Of course, we assume for the special case of the end-of-word marker that $Pi(])=Pc(])=0$, thus preventing the PFSM from generating this marker. Let us first assume that each segment $X(j)$, $1 \leq j \leq M$ contains a unique phonemic label $x(j)$. It can be shown ([5]) that the probability that Y has been pronounced is the probability that the PFSM associated to Y produces $x(1).x(2)\ldots x(P)$ when transiting from $S(0)$ to $S(M+1)$. Let us denote by $L(i,j)$ the probability for the PFSM to enter $S(i)$ after having produced $x(1).x(2)\ldots x(j)$. Then we have:

(1) $L(i,j) = Lc(i,j) + Li(i,j) + Lo(i,j)$ where

(2) $Lc(i,j) = L(i-1,j-1) \times Pc(y(i)) \times qc(x(j)/y(i))$

(3) $Li(i,j) = L(i,j-1) \times Pi(y(i+1)) \times qi(x(j)/y(i+1))$

(4) $Lo(i,j) = L(i-1,j) \times Po(y(i))$

Lc(i,j), Li(i,j) and Lo(i,j) denote respectively the probability for the PFSM to enter state S(i) after confusing x(j) and y(i), inserting x(i) or missing y(i). Equations (1), (2), (3) and (4) are valid when $1 \leq i \leq M$ and $1 \leq j \leq P$. If we add the following conditions:

(2.1) $(\forall i)(1 \leq i \leq M)$ Lc(i,0)=0, $(\forall j)(1 \leq j \leq P)$ Lc(0,j)=0
Lc(0,0)=1

and

(3.1) $(\forall i)(0 \leq i \leq M)$ Li(i,0)=Lo(i,0)=0
$(\forall j)(0 \leq j \leq P)$ Li(0,j)=Lo(0,j)=0

(1), (2), (3) and (4) still remain valid for $0 \leq i \leq M$ and $0 \leq j \leq P$.

Finally, the recursion scheme is completed by

(5) $(\forall j)(0 \leq j \leq P)$ L(M+1,j) = Lo(M+1,j) = L(M,j)

which derives from the fact that Pc()=0, Po()=1 and that S(M+1) has no insertion edge.

Equations (2) and (3) need to be modified in order to deal with the case of segments containing multiple phonetic labels. There is no loss of generality in assuming that every segment X(j) contains an equal number n of phonetic labels. X(j) is then composed of a vector $\{x(j,1), \ldots, x(j,n)\}$ of n phonetic labels and of a vector of probability $\{p(j,1), \ldots, p(j,n)\}$. Equations (2) and (3) are modified by replacing the terms qc(x/y) and qi(x/y) by their mean value over X(j), thus having:

(6) $Lc(i,j) = L(i-1,j-1) \times Pc(y(i)) \times \sum_{k=1}^{n} p(j,k) \times qc(x(j,k)/y(i))$

(7) $Li(i,j) = L(i,j-1) \times Pi(y(i+1)) \times \sum_{k=1}^{n} p(j,k) \times qi(x(j,k)/y(i+1))$

2.3. Application of the method to the detection of words in a sentence

Let V = {Y(1), ..., Y(d)} be a d-words vocabulary to be recognized and X = (X(1), ... , X(N)) be the phonetic translation of a pronounced sentence. The problem consists in comparing every word Y of V to each subsequence (X(k+1), ... ,X(k+P)) of X (0<k<k+P<N). Every such comparison provides a likelihood L. If L is a "satisfying" value, (for example, if L reaches a given threshold), then Y is said to be detected between segments X(k+1) and X(k+P). Otherwise, no output is produced.

3. DESCRIPTION OF THE NETWORK

3.1. The network

The basic design idea consists in attaching a processor to every point i,j. Let us denote by <i,j> the processor attached to point i,j. <i,j> receives the values Lc(i,j), Lo(i,j) and Li(i,j). It first computes L(i,j) by applying equation (1), and then computes Lc(i+1,j+1), Li(i,j+1) and Lo(i+1,j) by applying equations (5), (6) and (4) respectively. In order to compute Lc(i+1,j+1) and Li(i,j+1) <i,j> must receive X(j) and y(i).

The basic processor configuration and the network structure are direct consequences of the above considerations. The network is made of (M+1)x(P+1) processors labelled <i,j> with $0\leq i\leq M$ and $0\leq j\leq P$. <i,j> has to be connected to <i-1,j-1>, <i-1,j> and <i,j-1> in order to compute L(i,j). Consequently, the network has the structure displayed in Figure 2.

Every processor possesses six input/output ports (3 input ports and 3 output ports). Connection lines between processors are unidirectionnal (their direction is indicated by arrows). We suppose that interprocessor communications obey a producer-consumer scheme. For a given processor, Input/output ports are denoted as follows: Id (diagonal Input), Iv (vertical Input), Ih (horizontal Input) for Inputs and Od (diagonal Output), Ov (vertical Output) and Oh (horizontal Output) for Outputs.

3.2. Mode of operation of the network

3.2.1. Basic cycle of the processor

Consider the processor <i,j>, and suppose that the following data are avalaible as inputs :
- on Iv, successively X(j+1) and Lo(i,j),
- on Id, Lc(i,j),
- on Ih, successively y(i+1) and Li(i,j).

During its basic cycle, the processor computes L(i,j) by applying equation (1) and computes Li(i,j+1), Lc(i+1,j+1) and Lo(i+1,j) by applying respectively equations (7), (6) and (4). This means that the processor memorizes permanently the probability distributions qc(x/y), qi(x,y) Pi(y), Pc(y) and Po(y). Processor <i,j> then outputs:
- successively y(i+1) and Li(i,j+1) on Oh,
- Lc(i+1,j+1) on Od,
- successively X(j+1) and Lo(i+1,j) on Ov.

This computation is valid for the inner nodes, that is

the nodes i,j satisfying $1 \le i \le M$ and $1 \le j \le P$. The following remarks show that it is still valid for the nodes on the boundaries.

Note that equations (2.1) and (3.1) define the initial values to be sent to processors <i,j> where i=0 and $1 \le j \le P-1$ or j=0 and $1 \le i \le M$. Provided that these values are correctly sent on ports Iv, Id, and Ih, the above program does correctly the computation associated to these nodes.

Note also that processors <M,j> where $1 \le j \le P-1$ produce L(M+1,j) on their Ov port, due to the fact that Pc(])=0.

Finally, consider processors <i,P> where $0 \le i \le M$. In order for such processors to behave exactly as the others, it is convenient to introduce a fictitious end-of-sentence segment X(P+1) consisting in the unique end-of-word marker]. If we set

(8) $(\forall y)\ qc(]/y)=qi(]/y)=0$

it can be seen that <i,P> delivers Lo(i+1,P) on its Ov port, and consequently that <M,P> delivers L(M+1,P)=Lo(M,P) on its Ov output port.

3.2.2. Detection of a word

Suppose that the network is synchronous. At times 0, 1, ... ,t, ... each processor executes the above computation. We want to use our network in order to detect the word Y = (y(1), y(2), ... , y(M)) in the sequence X = (X(1), ... , X(P)). Figure 3 illustrates the functioning of the network.

At time t=0, <0,0> receives X(1) and 0 on Iv, 1 on Id and y(1) and 0 on Ih. Applying the basic cycle program, <0,0> outputs X(1) and Lo(1,0) on Ov, Lc(1,1) on Od and y(1) followed by Li(0,1) on Oh.

At time t=1, <1,0> receives the values sent by <0,0> on Iv, 0 on Id, y(2) and 0 on Ih. It thus outputs X(1) and Lo(2,0) on Ov, Lc(2,1) on Od, y(2) and Li(1,1) on Oh. At the same time, <0,1> receives X(2) and 0 on Iv and 0 on Id. It then outputs X(2) and Lo(1,1) on Ov, Lc(1,2) on Od, y(1) and Li(0,2) on Oh.

More generally, at time t, X(t+1) and 0 are made available on Iv of processor <0,t> and 0 on Id of <0,t>. In the same manner, y(t+1) and 0 are sent on Ih of <t,0> and 0 on Id of <t,0>. Then the diagonal of processors <i,j> where i+j=t delivers X(i+1) and Lo(i+1,j) on their output port Ov, Lc(i+1,i+1) on Od and y(i+1) followed by Li(i,j+1) on Oh.

One can see that the probability L(M+1,P) of the

matching between Y and X appears on processor <M,P> at time t=M+P. More generally, the probability L(M,j) of the matching of Y to the sequence (X(1), ... , X(j)) (j≤P) is given by the output port Ov of the processor <M,j> at time t=M+j. Let us emphasize that the above network still works for the detection of words of length l<M, provided they are completed with M-1 end-of-word markers]. It may also be applied for detection of words in strings of length l<P, provides they are completed with P-1 end-of-sentence markers.

3.2.3 Pilelined mode of operation of the network
Assume that all the words of the vocabulary V have a length less than or equal to M. We have seen that at time t, only the processors <i,j> (i+j=t) are active. This suggests a more optimized method for introducing data into the network. We present two mode of pipelining: pipelining the inputs y(i) (referred to as pipelining on the words) and pipelining the inputs X(j) (referred to as pipelining on the sentence).

3.2.3.a Pipelining on the words
By fixing X(1), ... , X(P), and presenting successives words Y(∅), Y(ℓ), Y(d) of V, it is possible to consecutively match every word of V against the same subsequence of the sentence. Let us denote Y(ℓ)=y(ℓ,1).y(ℓ,2). ... .y(ℓ,M) the word Y(ℓ) of V. Data are presented to the network in the following way.   Processor <i,∅> receives at time t (t≥i) the value y(t,i+1) on its Ih port. <∅,j> receives at time t (t≥j) the value X(j+1). At time t (t≥M+j), processor <M,j> outputs the probability of the matching between Y(t-M-j) and X(1), ... ,X(j).

3.2.3.b Pipelining on the sentence
By fixing the word Y=y(1). ... y(M) and shifting segments of the sentence from left to right, it is possible to successively match every subsequence of X(1), ... , X(P) to Y. In this mode of functionning, data are inserted as follows. Processor <i,∅> receives at time t (t≥i) the phoneme y(i+1). Processor <∅,j> receives at time t (t≥j) the value X(t+1) if t≤P or {]} if t>P. At time t (t≥M+j), processor <M,j> outputs the probability of matching Y against X(k+1), ... , X(k+j) where k=t-M-j.

3.2.3.c Combining the two pipelining modes of operation
The most natural way to solve the whole word-spotting problem is to compare all the words of the vocabulary

successively to subsequences starting at X(1) then to subsequences starting at X(2) and so on. This mode of operation allows to process the sentence on a phoneme-by-phoneme basis i.e. from the left to the right as soon as phonemes are produced by the phonemic analyzer.

4 CONCLUSION

In this paper, we have shown how the combined use of pipelining and multiprocessing may be applied to the detection of words in continuous speech. Let us emphasize two characteristics of our network:
- processors are identical,
- inter-processor connexions are very regular.

These observations suggest an implementation of this machine using VLSI techniques. This aspect is the object of our present investigations.

A quantitative evaluation of the expected performances of this network has been carried out. It shows that this method may be used for the real time detection of large vocabularies (1000 words), due to simplicity of the computation executed at each cycle.

BIBLIOGRAPHY

[1] Trends in Speech Recognition, W.A. LEA Editor, Prentice-Hall, 1980.
[2] C. Mead, L. Conway, Introduction to VLSI systems, Addison-Wesley, 1980.
[3] R. Bellman, S. Dreyfus, Applied Dynamic Programming, New Jersey, Princeton University Press, 1962.
[4] L. Buisson, G. Mercier, J.Y. Gresser, M. Querre, R.Vives, Phonetic Decoding for Automatic Recognition of Words, Speech Communication Seminar, Stokholm, August 1974.
[5] L.R. Bahl, F. Jelinek, Decoding for Channels with Insertions, Deletions and substitutions with applications to Speech Recognition, IEEE Trans on Information Theory, 21, number 4, july 1976, pp 404-411.
[6] H. Sakoe and S. Chiba, Dynamic Programming Algorithm Optimization for Spoken Word Recognition, IEEE Trans. on ASSP, vol. ASSP-26, number 1, february 1978, pp 43-49.
[7] G. Mercier, A. Nouhen, P. Quinton, J. Siroux, The Keal Speech Understanding System, in Spoken Language Generation and Understanding, J.C. Simon Editor, Proceedings of the NATO ASI, Bonas, 1980.

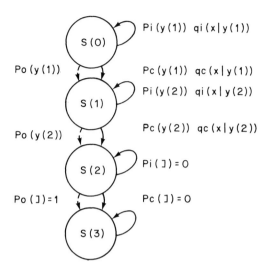

Figure 1: The PFSM associated with the word y(1).y(2).]

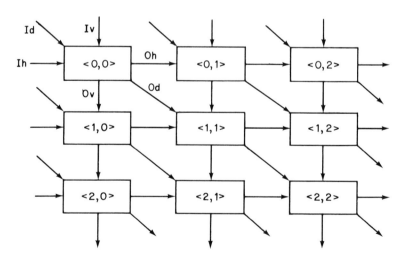

Figure 2: Structure of the network.

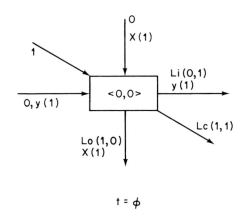
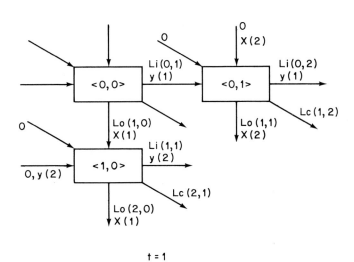

Figure 3: Functioning of the network.

CARRY-SAVE ARRAYS FOR VLSI SIGNAL PROCESSING

P.B. Denyer* and D.J. Myers†

*Department of Electrical Engineering, University of
Edinburgh, Edinburgh EH9 3JL, UK.

†British Telecom Research Laboratories, Martlesham Heath,
Ipswich, IP5 7RE, UK.

INTRODUCTION

It is well known that the parallel multiply function may be implemented in a regular array of carry-save adders. In this paper we show an extension of the concept to very large carry-save arrays that implement complete filter functions on high-speed parallel data. The arrays are computationally efficient and devoid of the communication overheads normally associated with parallel systems. They are therefore extremely suitable for VLSI implementation.

PROCESSOR DEFINITION

We present in Fig.1 a linear network for the solution of vector-vector and vector-matrix problems, similar to that proposed by Kung and Leiserson (1980). The network is a cascade of inner-product step computers, which each realise the process:

$$c' \leftarrow c + ab \qquad (1)$$

where a delay of one sample period (denoted Δ in the diagram) is implied between the step inputs a,b,c and the new output c'. Note that, unlike the architectures proposed by Kung and Leiserson, the input data is here presented simultaneously to all processors in the network. Although global communication paths are generally undesirable, this feature results in more efficient solutions to several real-time signal processing problems.

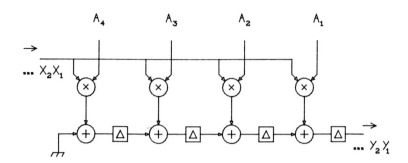

Fig.1(a) Linear network for convolution and transversal filtering.

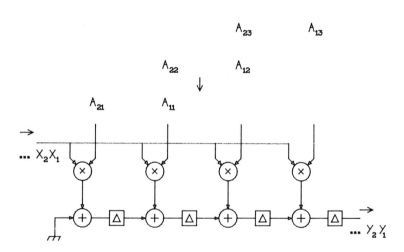

Fig.1(b) Linear network for matrix-vector product formation.

For example, the convolution of two vectors $\underline{x}$ and $\underline{a}$ may be implemented as in Fig.1(a), by preloading and latching one of the vectors as the set of multiplier coefficients over N processors. The output sequence is then:

$$y_i = \sum_{n=1}^{N} x_{i-n} a_n, \qquad (2)$$

which is the convolution function desired.

If the coefficients are latched and stationary and the input vector $\underline{x}$ is a continuous sampled signal, then the network becomes a form of Finite Impulse Response or Transversal Filter which is useful for linear-phase frequency filtering and for various matched filtering functions.

Similarly, the network may be used to compute matrix-vector products by presenting the matrix coefficients to the network as in Fig.1(b). The output sequence

$$y_i = \sum_{n=1}^{N} x_n a_{in} \qquad (3)$$

is obtained, which forms a solution vector every N sample periods, using a linear array of (M+N-1) processors for a matrix of dimension MxN. This type of filter is a useful tool for all types of discrete data transformation, e.g. DFT.

Direct implementation of the networks shown in Fig.1 leads to an inefficient circuit topology. Most significantly there is a predominance of parallel data bussing, and the parallel adders in the accumulation chain may limit the sample rate. Recognising that pipelining within these adders (and also within the multipliers) is the key to optimising the data rate, we have investigated pipelined multiply-add structures for inner-product step realisation. Although several of these are of interest, we report here a single structure that seems to offer the most elegant VLSI realisation in terms of its regularity and flexibility in use.

PROCESSOR REALISATION

The Bit-Plane

It is possible to represent any multiplier coefficient a_n as a weighted sum of m binary integers:

$$a_n = -2^0 a_{n,m} + 2^{-1} a_{n,m-1} + \ldots 2^{-(m-1)} a_{n,1} \qquad (4)$$

where $a_{n,x} = 1$ or 0, and m is the coefficient word length. The convolution equation(2) may therefore be represented as

$$y_i = \sum_{n=1}^{N} x_{i-n} (-2^0 a_{n,m} + 2^{-1} a_{n,m-1} + \ldots 2^{-(m-1)} a_{n,1})$$

$$= -2^0 \sum_{n=1}^{N} x_{i-n} a_{n,m} + \ldots 2^{-(m-1)} \sum_{n=1}^{N} x_{i-n} a_{n,1}. \quad (5)$$

Eqn.(5) states that a complete filter can be realised as a weighted sum of the outputs of m binary-coefficient filters or, as we shall label them, bit-planes. It is easily shown that the matrix-vector operation may be represented in a similar manner.

The advantage of the bit-plane approach to VLSI filter construction lies in the efficient circuit topologies with which it may be implemented.

Bit-Plane Realisation

We wish to realise a network of inner-product step processors in which the multiplier coefficients have single bit representations. Under these circumstances the multipliers reduce to AND gates, and the network can be realised as a clocked cascade of parallel adder registers, as shown in Fig.2(a).

In such a realisation potential speed (or area) problems arise in the generation and propagation of carry signals within the parallel adders. These may be resolved by horizontally pipelining the carry process, so that carries are saved and passed across the array at the same rate as sums are passed down the array, as shown in Fig.2(b) in which the individual cells represent clocked carry-save adders, including the product-forming AND gates. The data rate achievable with this form of pipelining is extremely high since the clock rate is limited only by the delay through a single adder.

Now the horizontal pipeline process skews data within the array. This is illustrated in Fig.2(b) by the diagonal lines that mark 'computational wavefronts' as data passes through the array. In effect each vertical column processes data in a time frame that is separated from its neighbours by one sample period. This does not alter the function of the array, but some preskewing and deskewing interfaces are required at the filter input and output respectively. These take the form of time-delay wedges which are simple to implement and represent an acceptable overhead.

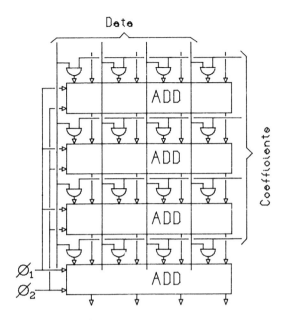

Fig.2(a) Register level bit-plane realisation

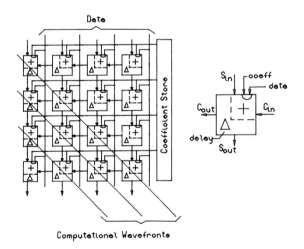

Fig.2(b) Cell level bit-plane realisation with horizontal pipelining.

A further consequence of horizontal pipelining occurs in applications for which the coefficients are time-variable, e.g. the matrix-vector product. In these cases the coefficients must also be pipelined across the array to synchronise with carry and data bits.

Not shown in these figures are the coefficient stores. These may be elementary latches for coefficient vectors, or recirculating shift registers for coefficient matrices. In either case the store may be loaded serially through one pin, and should be located close to the array to avoid interconnection problems.

Processor Realisation

The complete processor function is realised by weighting and adding together the outputs of m bit-planes, where m is the desired coefficient word-length. A pipelined tree of adders might be used, but this results in excessive interconnection, and where large bit-planes occupy single chips can cause severe pin- and package-count problems. A more elegant solution is to cascade the bit-planes noting that a free accumulator input is available at the top of each bit-plane, and that planes may physically abut with no data-routing overhead.

An example of this technique is illustrated in Fig.3, which shows a detailed floor-plan for a hypothetical FIR filter realisation with:

>   Filter length : 6 processors
>   Input word length : 8 bits
> Coefficient word length : 4 bits

The filter is formed by cascading four bit-planes, each six registers deep. The width of the array is appropriately increased as word lengths increase within the filter.

Binary weighting of the bit-plane outputs is achieved by a single-bit shift of the input data between planes, which involves no area penalty. The most significant bit-plane may be arranged to achieve a subtraction of the input data for systems using 2's complement arithmetic. It is interesting to note that, by starting first with the most significant bit-plane and proceeding down the cascade to the least significant, the outputs of the bit-planes are increasingly accurate estimates of the eventual filter output.

When cascading bit-planes it is important that the input data be correctly synchronised with the accumulating outputs.

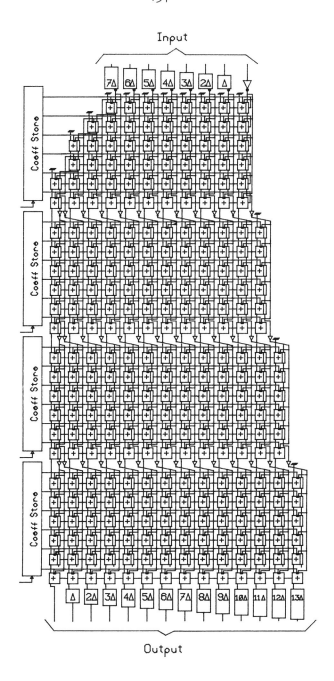

Fig.3   Array floor-plan for an FIR filter.

For this purpose the cascaded and shifted input data must be delayed by (N-1) samples between planes, which is achieved by appending single delay stages to appropriate cells. These are visible in Fig.3 along with buffers to drive the data into the following planes. To complete the processor realisation, skewing and deskewing time wedges are added at the data input and at the filter output.

The topological efficiency of this type of filter construction is remarkable. Computation progresses evenly along a data path which is always locally available, i.e. there are no parallel routing overheads. Also, the array is composed of identical cells which may operate at a pipeline clock rate limited only by their own performance, so that the computational power of the array is maximised.

It is not difficult to define a structure like that of Fig.3 to suit any processor requirement in terms of filter and word length, indeed the problem is well suited to a 'silicon-compiler' solution. However, even at VLSI circuit densities it may not always be possible to include a filter of the desired proportions on a single integrated circuit. In these cases filters may be cascaded in a similar manner to bit-planes, that is by routing the output of one filter into the top of the next. For this purpose it may become necessary to extend the array dimension to accommodate increasing word lengths through the cascade of filters. An alternative solution to this problem for very long filters is to realise a single large VLSI bit-plane. Complete filters may be constructed by cascading these devices. The advantage of this form as a standard product lies in the memory-like regularity of the large bit-plane. Each row of the array is identical, so some form of redundancy might be included to enhance yield, for example by using a fuse technology (Eaton *et al* 1981).

DESIGN NOTES

The concepts that we have presented have thus far not been restricted to any particular technology; in this section we give some indication of the size and performance that may be expected of processor realisations, at least in one particular VLSI technology.

We have investigated the design of these arrays in silicon-gate NMOS technology. A suitable clocked carry-save adder and data delay circuit is shown in Fig.4. Other realisations are possible, but this circuit offers a good compromise between layout area and speed. At process

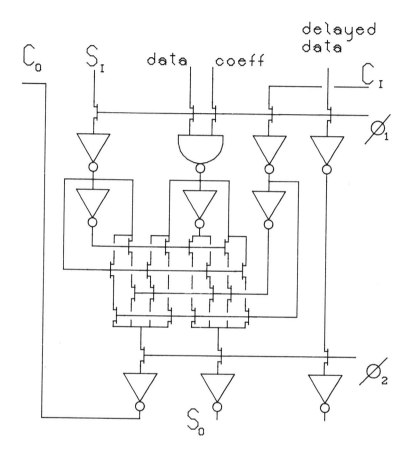

Fig.4 NMOS realisation of array cell.

feature sizes of 5μm the cell occupies approximately 250μm x 200μm and dissipates 1.5mW at clock rates up to 4MHz. Thus array dimensions up to 16 x 16 cells are feasible within die areas less than 20mm^2, and total dissipation around 400mW. Projection of this data to VLSI geometries shows that at feature sizes of 2μm, arrays containing 2000 cells at data rates of 10MHz are viable within the same die-area and power dissipation limits. For more demanding applications these conservative estimates are linearly improved with increases in allowable die size and power dissipation.

Regarding pin-count, entirely integrated filters pose little problem, where only data input and filter output signals need be interfaced. Where filters or bit-planes are to be cascaded the pin-count may become restrictive, however a cascadable bit-plane of any length operating on 8 bit data, with 20 accumulation bits passed between planes is viable within 64 pins.

CONCLUSIONS

We have shown how vector computation problems, such as FIR filtering and discrete transformation, may be implemented in large arrays of pipelined carry-save adders. Two features of such processor realisations are of note. Firstly, no space is wasted in runs of parallel interconnection - the data flows uniformly and unidirectionally through a regular computational array. Secondly, the performance of the array is optimal in the sense that the pipeline data rate is limited only by the processing rate of the common cell, and all cells operate at 100% duty cycle. Together these features ensure that computational performance is maximised within the available die area.

REFERENCES

Kung, H.T. and Leiserson, C.E.(1980). Algorithms for VLSI Processor Arrays. *In* "Introduction to VLSI Systems" (Mead and Conway). pp271-292. Addison Wesley.

Eaton, S.S. Wottan, D., Slemmer, W. and Brady, J. (1981). A 100ns 64K Dynamic RAM using Redundancy Techniques. *Digest 1981 IEEE International Solid-State Circuits Conference*, 84-85.

SESSION 4

REST
A Leaf Cell Design System

R. C. Mosteller

Department of Computer Science
California Institute of Technology
Pasadena, California

INTRODUCTION

We are now in a new revolution, the microelectronic revolution (Noyce 77). This microelectronic revolution has provided the means to put 50,000 or more circuit elements on a single chip. As each year goes by, semiconductor manufacturers are able to put an ever increasing number of circuit elements on a single chip. With the increase in circuit complexity of a VLSI chip, the design time also increases at a rapid rate(Moore 79). In order to cope with the larger complexity and design time, a structured design methodoly must be adopted.

Structured integrated circuit design is similar to the structured design of programs. The design is partitioned into small manageable pieces called cells which are similar to procedures in programming languages. A cell is considered to be rectangular in shape as shown in figure 1. All geometrical data is on the interior of the rectangle and considered to be the property of the cell. Each cell is composed of other cells or primitive elements which form a hierarchy of cells akin to the nesting of procedures. The number of elements in a cell is kept small, approximately seven, corresponding to the average person's short term memory limitation(Miller 1956). In structured programming it is desirable to have no global variables and likewise in structured design global wires are undesirable. A global wire is one that runs across several cells and is not part of their definition. That is, wires are not laid on top of cells, they must be part of the cell definition. Connection to the cells occur at the perimeter of the cell at locations called ports. Thus, cells are connected by abutment where possible. This structured approach promotes a regular structure with a consistent wiring strategy. This design style for VLSI systems is presented in Mead and Conway(1980).

The cells of a structured design are partitioned into two types: composition cells and leaf cells. A composition cell is a cell which contains other composition cells or leaf cells, while a leaf cell contains only wiring and primitive circuit elements. This partitioning of cell types is called a separated hierarchy(Rowson 1980) as shown in figure 2. Note that the leaf cells are at the bottom of the tree while the composition cells form the branches. Each cell, being either a composition cell or a leaf cell, contains a small number of elements that are easily grasped by a designer. Therefore a designer can comfortably work on a single cell in the hierarchy of his chip. This also provides the ability to partition the design among many designers.

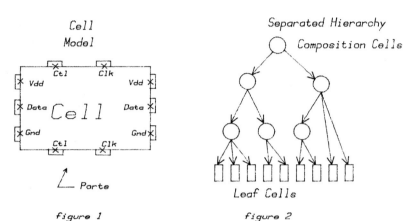

figure 1 figure 2

There are two major divisions in a computer aided design system consistent with this design philosophy. A leaf cell design system called REST and a composition tool for constructing the VLSI chip from leaf cells and composition cells. This computer aided design system is shown in figure 3. This system is a simple design system with the interfaces between parts using standard text files. With this system, it is not necessary nor desirable to have an exotic data base system. What is required is a flexible file system.

The essential task in leaf cell design is to capture both the topology and layout in the description of each cell. This is best accomplished by a sticks notation devised by Williams(1977) and described in Mead and Conway(1980). The REST system provides the means to process the stick notation, translate this notation to a abstract stick representation, and a sticks compaction tool. The interface with the rest of the system is through a simple disk file using a standard form called the Sticks Standard(Trimberger 1980). This standard provides a clean interface between the REST system and any composition tool. The testing of the leaf cells is done through a circuit simulator. This simulator accepts the Sticks Standard.

The task of assembling the chip is performed by a composition tool. The input to the composition tool are the leaf cells from REST and composition cells defining the spatial positioning, structure and

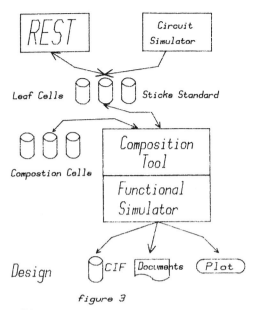

figure 3

behavioral data. The description of the composition is best accomplished in a textual manner with a composition language as in the SPAM Language(Segal 80). The main point here is that the spatial positioning, structural and behavior data is captured in a simple manner at one place in the design description.

The actual composition or tiling of the VLSI chip is performed by aligning the pitches of the cells or by adding wiring cells. Alignment is performed by stretching a cell to to fit its environment. The stretching is performed by the composition tool since detailed knowledge of the geometry is not necessary in the stretching process. Compaction could also be used in the alignment process by adding constraints to the leaf cells and using REST to process the cell.

The functional testing of the chip is done with the SPAM system(Segal 1980). In the SPAM system the composition cells also contain the behavioral description in addition to spatial positioning information and structural data. SPAM is a multi valued functional simulator capable of simulating at any level of the hierarchy.

The output of the composition tool is a design description in text format known as the Caltech Intermediate Form (CIF)(Mead 1980). It is also possible to request various plots and some additional documentation. CIF can be processed by various silicon foundry tools to produce mask geometry.

REST

The topological, structural and geometric data for leaf cells is captured by the REST system. The input is a simple colored sketch of lines and boxes representing a circuit shown in figure 4. This colored sketch is digested by the REST process which produces a compacted sticks representation. These two functions lead to a natural partitioning of tasks into two parts: 1) The line and box diagramming system that run in a local color graphics station and 2) the main program that does the major processing. At the present time, REST permits only orthogonal line drawings.

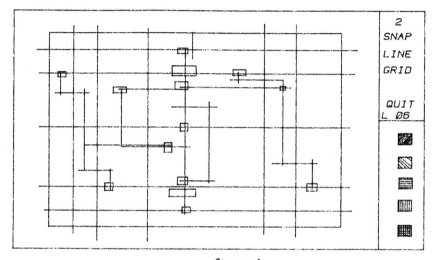

figure 4

The diagramming system known as the box editor runs in a color graphics display station called the Charles graphic station(Minter 1980). This station is composed of a video monitor, a DEC LSI-11 processor, a frame buffer, a DEC VT52 terminal, and a Hewlett Packard four pen color plotter. There are 4 bits per pixel on the color monitor which are memory mapped into the address space of the LSI 11. This work station is linked to a host computer via a data communication line at 9600 baud. The graphical entry is via a three button mouse. The commands to the box editor were kept simple and small in number to provide a good human interface.

The box editor has three commands, which map directly to the mouse buttons, and two modes. The commands are: line/box add, line/box delete and line/box move. The modes are line and box. Color selection, mode selection, and termination are accomplished via a menu as shown in figure 4. This provides an easily remembered set of commands and a reasonable human interface.

The advantages of separating the graphical editing function from the main program is twofold. First graphic editing requires fast computer response so the designer can achieve true interactive drawing capabilities. This fast response is accomplished by using a dedicated local processor for the graphic editing. Second, many different graphic work stations, with their own local editors may be used. REST can currently interface two graphic work stations, the Charles terminal and the DEC GIGI.

A sample input sketch for REST is shown in figure 4. The important characteristic of the sketch is that it is a rough drawing. There is no need to be neat in the sketch since the REST processor will recognize connection points as any two similarly colored lines which are in close proximity, transistors as red lines crossing green, and contacts as gray boxes over colored lines. A designer generally would sketch out his sticks drawing on the Charles terminal as if he were doing his stick drawing on paper. However, the lines are easier to erase on the screen than they are on paper. Blue lines are recognized as metal, red lines are recognized as polysilicon, and green lines are recognized as diffusion. The meaning of the colors are consistent with Mead and Conway(1980). Internally, these colors are derived from a table so the meaning may be changed easily for color preference or for different technologies.

The interpretation of the sticks drawing is done by the main program of REST which runs on the host computer. The line drawing interpreter analyzes the crude line drawing for connections between lines, simple devices, connection points, and contacts. In the drawing in figure 4 an enhancement device is recognized as red crossing green, a depletion device is recognized as red crossing green with a yellow box on top and a pullup is recognized as a red line on top of a green with a grey box at one end of the red and a yellow box in close proximity. Contacts are recognized as grey boxes over lines. The device ratios and line widths are taken from a user defined table. In addition, the line drawing interpreter correlates the newly interpreted drawing .with the previous one, if it exists. This operation is done so that old component names, device ratios, and line widths can be transferred to the new sticks representation for the cell.

The recognition of the sketch in figure 4 is shown in figure 5. This stick cell is shown in the standard sticks diagram form that REST generates. In fact, this drawing was plotted with the attached Hewlett Packard 7721 plotter. In the drawing, lines that were close, and of like colors, were snapped together. Lines that overlapped were trimmed off. The drawing shows transistors fleshed out as boxes

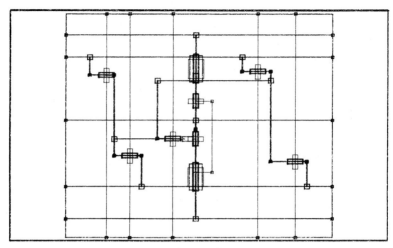

figure 5

with stylized lines representing the wires. The sizes of the transistors are proportional to their ratios.

The REST system follows the principle of isomorphism. The relationship between the line drawing and the internal abstract sticks representation is isomorphic. This allows fast switching from editing to REST proper. The abstract stick representation is also isomorphic to the physical representation. This provides a consistent transformation from sticks representation to physical representation.

The interpreted cell in figure 5 is shown compacted in figure 6 and in its physical form. Since the sticks representation is isomorphic, it could just as easily have been shown in sticks form. Compaction in REST is accomplished by compression in the vertical axis followed by compression in the horizontal axis. This order can be user controlled. REST allows lines to jog in either direction at wire intersection points. Topological features are allowed to cross provided they are not constrained by design rules. Constraints defined by the user are also allowed. REST employs a table driven compaction algorithm using a table that can easily be changed for various design rules. Currently REST uses the design rules in Mead and Conway(1980).

A graph based compaction algorithm is used in REST. The use of a graph to do compaction is not new, this method has been used by Williams(1979), Hsueh(1979), and others. A pictorial representation of the graph for a shift register cell is shown in figure 7. This example is for vertical compaction. Each node in the graph represents a group. A group is a set of primitive elements that moves as a unit. Node 1 and 7 in figure 7 represent the bottom and top edge of the cell. Node 2 is the metal wire with an attached connector. Each additional

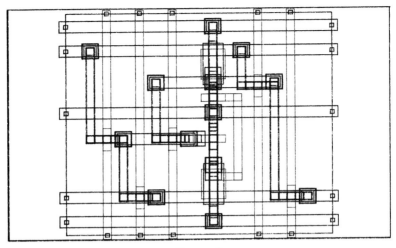

figure 6

node represents a group in the graph. The node number in the graph has a corresponding number in the sticks representation next to the nodes group. The determination of the groups is a simple mapping from REST internal sticks data structure.

The branches of the graph represent the design rules. Each branch contains the minimum separation between its two nodes. The generation of the branches from the groups and sticks representation normally has a complexity of N-squared. This is due to the fact that each group must be compared to every other. In REST the complexity of branch determination is order N. This operation is accomplished by using a group shadow mechanism. During branch determination, groups will become partially covered by other groups. The portion of a group that is covered is subtracted from the group. After each comparison the group is decreased in size. When a group is completely covered the comparison for it will terminate.

Unlike other graph compaction methods, REST contains a weighted affinity factor in each branch. This factor is the amount of attractiveness between the two attached nodes. This factor is best explained by an example. For instance, an example cell has a metal wire running through the cell with a second metal line orthogonal to it and a diffusion strap crossing under the first metal wire. If compacted without an affinity factor, the diffusion strap will be elongated toward the edge of compacted cell. That is, the contact that connects the metal wires to the diffusion strap will fall to the edge when it is compacted or in some programs it would be centered between the first metal wire and edge of the cell. In REST, the elongation of the diffusion strap would not occur due to the use of the affinity factor. The diffusion wire would be kept as short as design

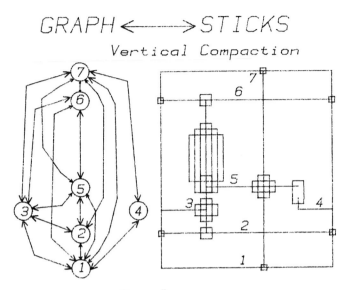

figure 7

rules permit. The branch for the diffusion strap would have a higher affinity than the metal line from the contact to the edge. The affinity in REST is table driven based on line type and line width. That is, the polysilicon and diffusion lines have greater affinity than metal. This factor can be altered based on design rules. The affinity factor controls the trade off in line length between the metal, diffusion and polysilicon wires.

In addition to the design rules and affinity factors the branches contain a maximum separation distance. This maximum is used for user defined constraints. The user constraint is of the form: COMP1+VALUE1 <= COMP2 <=COMP1+VALUE2. This user constraint allows absolute control over the compaction of the cell. The user constraints are generally used to compact a cell into an environment or to allow control over the connection ports.

The method used in solving the graph is straightforward and linear with time. The computer execution time for compaction is about the same time that it takes to read the cell from disk in the Sticks Standard to the internal representation. In REST, the response time is very fast. This allows the user to make several iterations of editing and compaction on a cell in a very short time.

REST provides a suite of utilities for displaying or plotting the cell, and for altering names, line widths, and device parameters. In addition, routines are provided to allow expansion of the cell. The

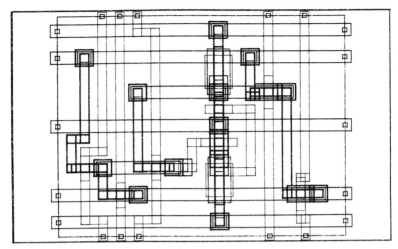
figure 8

edges of the cell can be extended to add additional logic. There is also a routine to expand about a point for logic insertion. Possible jog position points can also be added. In figure 8 the same cell is shown with line jogs added, line widths altered and annotation added. The vertical lines have been constrained to be at equal x values.

It is intended that REST provide a simple user interface with a few, yet powerful, commands. REST has an extensive help command so that query about any or all commands is possible. The commands are divided into four parts: edit, compaction, expansion, alteration and cell management. The edit command is used to start the box editor. The compaction commands are used to control compaction while the expansion commands are used to expand portions of the cell. The alteration commands are used to set device sizes, line widths and add constraints. The cell management commands are used to read and write the current cell to or from the Sticks Standard.

In normal use, a cell would be initially sketched with the box editor. This cell would then be interpreted and edited until the cell was logically correct. Then several iterations would be accomplished between compaction and editing until the desired degree of compaction is reached. The design of a leaf cell in this manner takes very little time. In fact the register cell for this paper was constructed in about ten minutes.

CONCLUSIONS

REST has been used at Caltech with great success to design the leaf cells for several large projects. One such project an eight bit stack data machine that used REST for several cells. A graphic processor chip GRIP used over 100 REST cells. Also a theorem proving chip employed mostly REST cells. This chip executes unification computation which is the main computing mechanism in the PROLOG system. This chip employed over 60 REST cells. A current project is the ETHERNET chip which has used over 50 REST cells. These projects are large in scope. REST substantially reduced the leaf cell design time for theses projects.

The Silicon Structures project at Caltech is using REST to define the structural idioms for several new VLSI designs. The current uses of REST have been for NMOS design. However, by changing the technology table, CMOS leaf cells could also be designed.

References

Hsueh, Min-Yu (1979). "Symbolic Layout and Compaction of Integrated Circuits" Memorandum No. UCB/ERL M79/80 University of California, Berkely

Mead, C.A. and Conway, L.A. (1980). "Introduction to VLSI Systems", Addison Wesley,1980

Miller, G.A. (1956). "The magical number seven, plus or minus two: some limits on our capacity for processing information" Psychology Review 1956 63, 81-97.

Minter, C. (1980). "Charles Terminal Care Package", California Institute of Technology, Silicon Structures Project,SSP MEMO #3804

Moore, G.E. (1979). "Are We Really Ready for VLSI?" Proceedings of the Caltech Conference on Very Large Scale Integration, 22-24 January 1979

Mosteller, R.C. (1980). "REST", California Institute of Technology Silicon Structures Project, SSP MEMO #4317

Noyce, R. (1977). "Microelectronics" A Scientific American Book Chapter 1. Microelectronics

Segal,R (1980). "SPAM", California Institute of Technology Silicon Structures Project, SSP MEMO #4029

Trimberger, S. (1980). "The Proposed Sticks Standard", California Institute of Technology Silicon Structures Project SSP MEMO #3487

Williams, J. (1977). "Sticks - A New Approach to LSI Design" M.S.E.E thesis, Dept. of Electrical Engineering and Computer Science M.I.T.

Williams, J. (1980). Private Communication with, John Williams

AN ALGEBRAIC APPROACH TO VLSI DESIGN

Luca Cardelli and Gordon Plotkin

Department of Computer Science, University of Edinburgh, Edinburgh EH9 3JZ, UK

1. INTRODUCTION

We propose an algebraic view of the hierarchical approach to VLSI design developed by Mead, Conway and others (Mead and Conway, 1980). VLSI networks are described by expressions of a many-sorted nMOS algebra, and the algebraic operators are designed to support a structural methodology. The algebra can be embedded in a programming language as an abstract data type (Goguen *et al.*, 1978), and since the emphasis is on expressions (they denote networks) it is natural to use an applicative language (Burge, 1975).

These ideas are really very general and should be useful wherever a structural approach is needed for graphical or geometrical information; in particular there is no difficulty with other technologies such as cMOS. The ideas largely originate with Milner (1979) and the small differences in the choice of operators are motivated by programming convenience and the fact that 1-1 connection is more natural for VLSI than many-many connection.

Our nMOS expressions will have various interpretations whether as graphs, geometric configurations, behaviours or other networks; in section 2 we give an informal one as pictures composed of coloured lines (*sticks*) and their intersections (*stones*). Each network will interface with its environment via a set of named coloured *ports* which determine its *sort*. The unary operators *renaming* and *restriction* manipulate the interface, and the more interesting binary operator *composition* joins smaller networks together into larger ones.

By adopting various abbreviations an extended notation is developed in section 3 which merges naturally into an applicative language. The control structures in the language provide conditional assemblies of networks and parameterizations, and a specialized iteration construct allows the convenient expression of the most common VLSI subsystems. We thus obtain what might be best called a high level chip assembly language.

The flexibility of parameterization of textual expressions has no graphical parallel and so we prefer the textual expression of networks. However, graphical tools should be used as much as possible especially for interactive visual feedback, and we believe that an algebraic approach might still be useful for hierarchical input and editing.

In conclusion, our algebraic notation can help the integration of the graphical and textual aspects of network design, because it can reflect the intended structure of networks and keep the connectivity information local and well-organized.

2. nMOS NETWORK ALGEBRAS

Our algebras are many-sorted, the idea being that the sort of a network determines its interface with its environment. We view networks as interfacing through a set of named ports (one name to each port), each on some layer. Formally, let {green,red,blue} be the set of *types* and let PNames be an infinite set of (*port*) *names* (and we use a,b to range over names and A,B,C to range over finite sets of names). Then a *sort* is a map s: A -> Types (and we put $|s| = A$); s(a) shows the layer of the port (named) a.

The elementary network components form the set, Γ, of *constants* of our algebra (and are ranged over by c); every constant, c, has sort $\sigma(c)$. Here is one reasonable choice for Γ giving the sort in an evident notation; Fig.1 pictures some of these elementary components.

```
Contacts
    Green        GCon:   {gn,gs,ge,gw:green}
    Red          RCon:   {rn,rs,re,rw:red}
    Blue         BCon:   {bn,bs,be,bw:blue}
    Green-Red    GRCon:  {gn,gs:green; re,rw:red}
    Red-Blue     RBCon:  {rn,rs:red; be,bw:blue}
    Blue-Green   BGCon:  {bn,bs:blue; ge,gw:green}
Transistors
    Enhancement  ETran:  {source,drain:green; gate,gate':red}
    Depletion    DTran:  {source,drain:green; gate,gate':red}
```

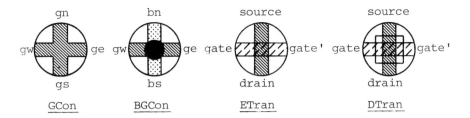

Fig.1 Some elementary components.

There are two unary operators for manipulating port names thereby changing the interface. The *restriction* operator, $\smallsetminus a$, removes the name a from the sort of a network; we make it postfix and often abbreviate $\smallsetminus a_1 \ldots \smallsetminus a_n$ to $\smallsetminus a_1, \ldots, a_n$. Fig.2 pictures DTran' = DTran$\smallsetminus$gate'. Formally for every s and a we have a unary operator $\smallsetminus a$: s -> s' (where $|s'|$ = $|s|\smallsetminus\{a\}$ and s'(a') = s(a') for a' in $|s'|$); thus if e is an algebraic expression of sort s then e$\smallsetminus$a is one of sort s'.

The *renaming* operator {r} (where r: A -> A' is a bijection) renames the ports of a network according to r; we make it postfix and often write $\{a_1\smallsetminus b_1, \ldots, a_n \smallsetminus b_n\}$ for r when $\{a_i\} \subseteq A$ and $b_i = r(a_i)$ and for a not in $\{a_i\}$ we have r(a) = a. Fig.2 pictures a power supply, ground and a butting-contact denoted, respectively, by the expressions,
   VDD = BGCon$\smallsetminus$gw {ge$\smallsetminus$high,bs$\smallsetminus$vdde,bn$\smallsetminus$vddw}   and
   GND = BGCon$\smallsetminus$ge {gw$\smallsetminus$low,bn$\smallsetminus$gnde,bs$\smallsetminus$gndw}   and
   BuCon= GRCon$\smallsetminus$gn,rw {gs$\smallsetminus$green,re$\smallsetminus$red}.
They have respective types VDD: {high:green; vdde,vddw:blue} and GND: {low:green; gnde,gndw:blue} and BuCon: {green:green; red:red}. Formally for every s and bijection r: $|s|$ -> A' we have a unary operator {r}: s -> s' where s' = s $\circ$ r^{-1}.

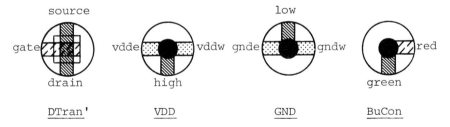

Fig.2 Some building-blocks.

The binary *composition* operator, [r] (where r: A -> A' is a bijection) composes two networks together according to r; the composition is allowed only if corresponding ports have

the same type and there are no two ports with the same name
among those not joined via r. We make [r] an infix,
associating to the left and often write it as $[a_1$--$b_1,\ldots,$
a_n--$b_n]$ where $\{a_i\}$ = A and $b_i = r(a_i)$. Fig.3 shows how to
make a resistor. First compose DTran" = DTran' {source⌢in}
with BuCon via [gate--red] obtaining DTran" [gate--red] BuCon
: {in,drain,green:green}; then compose the result with
GCon' = GCon⌢ge {gs⌢out} via [drain--gn,green--gw] obtaining
Res: {in,out:green} where
    Res = DTran" [gate--red] BuCon [drain--gn,green--gw] GCon'

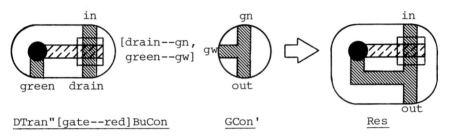

DTran"[gate--red]BuCon   GCon'   Res

Fig.3 Building a resistor.

Formally for every s,s' and bijection r: A -> A' where
A ⊆ |s| and A' ⊆ |s'| and s(a) = s'(r(a)) for every a in A
(type restriction) and such that B ∩ B' = ∅ where B = |s|∖A
and B' = |s'|∖A' (names refer to one port) we have a binary
operator [r]: s,s' -> s" where |s"| = B ∪ B' and s"(b) is
s(b) if b is in B and s'(b) otherwise.

As another example, an inverter of sort {in:red; out:green;
vdde,vddw,gnde,gndw:blue} can be built as
    Inv = (VDD[high--in]Res[out--gn](GCon⌢gw)[gs--source]
          ETran[drain--low]GND) {gate⌢in,ge⌢out}
Composition cells in the sense of Rowson (1980) can be
regarded as repeated applications of our composition with
some restriction, renaming and constants.

The pictures we have used above will now be formalised as
appropriate kinds of graphs; they can be regarded as stick
diagrams where connectivity is the only topological
information retained. Algebraically, what we have done above
is to give the *signature* of *n*MOS algebras and we now turn
to the most important algebra.

Definition An *n*MOS *network* is a quintuple N = <V,γ,A,π,E>
where V is a non-empty finite set and γ: V -> Γ (and we put
P = {<v,b>| b ∈ σ(γ(v))}) and A is a finite set of names and

π: A -> P is 1-1 and E ⊆ P × P subject to the following
conditions (where type(v,b) = σ(γ(v))(b)):
1. E is symmetric and a partial function.
2. If <v,b>E<v',b'> then type(v,b) = type(v',b') and v≠v'.
3. No <v,b> is both π(a) for some a and in the domain
   of E.
The *sort* of N is s where $|s| = A$ and $s(a) = type(\pi(a))$.

Intuitively, V is the set of vertices (or nodes) and γ(v)
is the the elementary component at v and A is the set of
port names and P is the set of ports and π(a) is the port
that a alone names and E is the connection relation between
ports. The first condition ensures that connection is
symmetric and any port is connected to at most one other;
the second ensures type-consistency and that there are no
self-loops; the third ensures no port is both named and
connected. We identify isomorphic networks when the
isomorphism is the identity on port names.

Now the *carriers* of the algebra are the networks of sort
s, for each s. Any constant, c, of sort s denotes the
network $<\{c\}, c \mapsto c, |s|, a \mapsto ><c,a>, \emptyset>$. The operations can be
given by simple definitions.
Restriction $<V,\gamma,A,\pi,E> \setminus a = <V,\gamma,A',a'+>_\pi a',E>$ where $A'=A\setminus\{a\}$.
Renaming $<V,\gamma,A,\pi,E>\{r\} = <V,\gamma,r(A),\pi \circ r^{-1},E>$
Composition $<V,\gamma,A,\pi,E>[r]<V',\gamma',A',\pi',E'> =$
$<V \cup V', \gamma \cup \gamma', B \cup B', \pi'', E \cup E' \cup E''>$ where we assume that
$V \cap V' = \emptyset$ and where $r: C \to C'$ and $B = A \setminus C$ and $B' = A' \setminus C'$
and $\pi''(b)$ is $\pi(b)$ if b is in B and $\pi'(b)$ otherwise and where
$E'' = \{<\pi(a),\pi'(r(a))>,<\pi'(r(a)),\pi(a)> \mid a \in C\}$.

It can easily be shown that every network is denoted by
some algebraic expressions, confirming the power of our
notation; also it can be decided in polynomial time whether
two expressions denote isomorphic networks, but we lack a
good upper bound. We can also axiomatise the equalities
between expressions by variants of Milner's laws of flow.
In the following x,y and z range over arbitrary expressions
such that the equations are between well-sorted expressions
of the same sort; we write σ(x) for the sort of x and
id_A: A -> A is the identity on A.
Composition $(x[id_A]y)[id_{A' \cup A''}]z = x[id_{A'' \cup A}](y[id_A,]z)$
$\qquad$ if $A'' \cap A = A \cap A' = A' \cap A'' = \emptyset$
$\qquad x[r' \circ r]y = x\{r \cup id_B\}[r']y$
$\qquad\qquad$ where $B =|\sigma(x) \setminus A$ where $r: A \to A'$
$\qquad x[r]y = y[r^{-1}]x$

Restriction $x \smallfrown a = x$ (if $a \notin |\sigma(x)|$)
            $x \smallfrown a \smallfrown b = x \smallfrown b \smallfrown a$
            $(x[r]y) \smallfrown a = (x \smallfrown a)[r](y \smallfrown a)$ (if $a \in |\sigma(x[r]y)|$)
Renaming $x\{id_A\} = x$ (where $A = |\sigma(x)|$)
            $x\{r\}\{r'\} = x\{r' \circ r\}$
            $x\{r\} \smallfrown b = (x \smallfrown a)\{r \smallfrown \{<a,b>\}\}$ (where $b = r(a)$)
            $(x[r]y)\{r' \cup r''\} = x\{r' \cup id_A\}[r]y\{id_{A'} \cup r''\}$
            (where $r: A \to A'$)

It can be shown that all these laws are true in the above
network algebra (consistency) and that any true equation in
the algebra can be proved from the laws (completeness). It
follows from this and the above remarks on the power of the
notation (definability) that our algebra is initial in the
class of algebras satisfying the laws. Any semantics for
nMOS networks should be such an algebra.

3. GEOMETRIC INTERPRETATION

Net Algebras can be embedded in a programming language by
adding an abstract data type *picture* together with some
special syntax for constants and operators. We illustrate
this process in the case of a geometric interpretation of
Net Algebras, obtaining a language capable of directly
expressing geometric layouts. The host language considered
here is Edinburgh ML (Gordon *et al.*, 1979) and the resulting
language is called Sticks & Stones. An implementation has
been carried out on the ERCC DEC-10 at Edinburgh University
(Cardelli, 1981) which makes interactive use of a colour
graphics display and also compiles pictures into CIF files.

In the geometric interpretation, a picture is a configuration of coloured geometric *figures* (generally rectangles
or polygons) and ports which are now named *vectors* with a
size and an orientation. We use *compound* port names like
g.east, a or a.b.1. Both figures and vectors have a fixed
displacement from a conventional origin which is local to
each picture; configurations are identified up to translation
and rotation. The geometric interpretation can be made formal,
but we concentrate here on other issues.

There is an infinite supply of constants, given by the
syntax exemplified in Fig.4. The elementary picture GCon
is a green box with lower left corner at 0^0 and upper right
corner at 2^2. It has four ports. For example, s is a green
port which is a vector starting from the point 0^0 and going
in direction 0 degrees counterclockwise from the x-axis for
a length of 2. Note that the names n (north), s (south) etc.

are only conventional; pictures have no predefined
orientation. The definition of constants requires precise
geometric information, and we would prefer graphical input
to the system.

```
    GCon = form (n: green port [2^2,180,2];
                 s: green port [0^0,   0,2];
                 e: green port [2^0,  90,2];
                 w: green port [0^2,270,2])
           with green box  [0^0,2^2].
    ETran= form (n: red   port [2^2,180,2];
                 s: red   port [0^0,   0,2];
                 e: green port [2^0,  90,2];
                 w: green port [0^2,270,2])
           with green box  [0^0,2^2]
           and  red   box  [0^0,2^2].
    DTran= form (n: red   port [3^4,180,2];
                 s: red   port [1^0,   0,2];
                 e: green port [4^1,  90,2];
                 w: green port [0^3,270,2])
           with green box  [0^1,4^3]
           and  red   box  [1^0,3^4]
           and  yellow box [0^0,4^4].
```

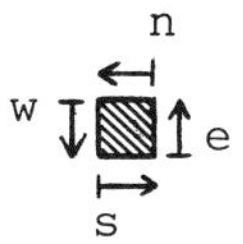

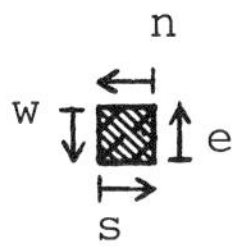

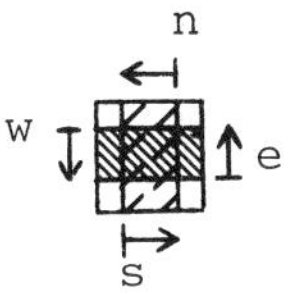

Fig.4 Some elementary geometric components.

Restriction and renaming operate in the way described in
section 2, but some abbreviations are introduced. There is
a *pattern matching* feature on compound port names. A name
like a.!.b.? represents all the names in a sort which match
the pattern, where ! matches any single atomic part and ?
matches a (possibly empty) list of atomic parts of a
compound name. The use of ? is restricted to the end of a
pattern to avoid ambiguities; the pattern ? matches all the
ports. Hence ⌐!.east means forget all the east ports and
{green.?⌐?} means rename all the green ports by dropping the
prefix green. Pattern matching is an abbreviation for the
appropriate enumeration of all the ports matching the pattern.

A renaming of the form {a⌐b move 2} is a case of *geometric
renaming*; the port a is renamed b and it is moved by 2
towards the east of the port (the north being the tip of
its vector). During this movement the port leaves a trail of
its passage, which is a polygon of the same colour as the
port. Geometric renaming is an abbreviation for the
composition of a suitable "trail" form. It is possible to
move, rotate and change the size of ports, and to compose
these actions.

A combination of pattern matching and geometric renaming is shown in Fig.5.

out = GCon⌢e {?⌢? move 2}
pos = ETrans {n⌢ctrl.n,s⌢ctrl.s,
              e⌢data.e,w⌢data.w} {?⌢? move 2}
neg = DTrans {n⌢ctrl.n,s⌢ctrl.s,
              e⌢data.e,w⌢data.w} {?⌢? move 1}

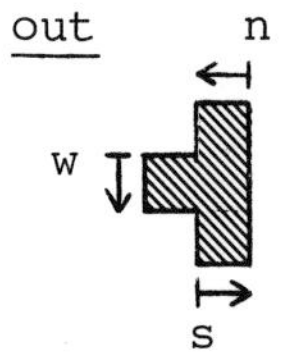

Fig.5 Some restrictions and renamings.

The T-shaped out is obtained from a green square GCon, first forgetting the e port, and then moving all the remaining ports outwards. The pictures pos and neg (not shown) have a cross shape.

Our final example (Fig.6 and 7) is a function taking a parameter n and producing a selector with n control inputs, n complemented control inputs and 2^n inputs. Some standard programming language features are used without explanation.

```
 1.  let sel n =
 2.  for i in 1::exp(2,n)
 3.  iter (for j in n::1
 4.        iter if bit(i-1,j-1) = 0
 5.              then pos[data.e--data.w]neg{ctrl.?⌢ctrl'.?}
 6.              else neg[data.e--data.w]pos{ctrl.?⌢ctrl'.?}
 7.        with [data.e--data.w])
 8.        [data.e.!--out.w] out
 9.  with [ctrl.s.!--ctrl.n.!,ctrl'.s.!--ctrl'.n.!,
            out.s--out.n]
10.  where rec bit(i,j) =
11.  if j=0 then i mod 2 else bit(i div 2,j-1)
```

Fig.6 A selector generator.

The circuit shown in Fig.7 is the result of the evaluation of sel 2 (selector with two control inputs). This selector program takes advantage of a specialized iteration construct:
   for <variable> in <list> iter <picture> with <connection>
where the <connection> is used between the <picture>'s produced as the iteration <variable> ranges through the <list> of values. Iteration also applies an automatic indexing appending the number n to all the ports produced at the n-th iteration (e.g. a becomes a.3), thus avoiding name clashes.

The selector is obtained by two nested iterations, first
building the rows and then joining them up into an array.
At the core of the double loop (lines 4-6) we have to choose
between a pair pos-neg' and a pair neg-pos' (where pos' and
neg' are pos and neg with their ctrl ports renamed to ctrl');
this is done using a function bit (defined in lines 10,11).
The inner loop (lines 3-7) connects all these pairs into a
row, with the variable j ranging from n to 1 (line 3). At
the end of the inner loop, an out element is added to the
right of the row (line 8). In the outer loop the variable i
ranges from 1 to 2^n (line 2) while all the rows are connected
from south to north (line 9). The exclamation marks in lines
8 and 9 take care of the indexes added to the ports during
the inner iterations.

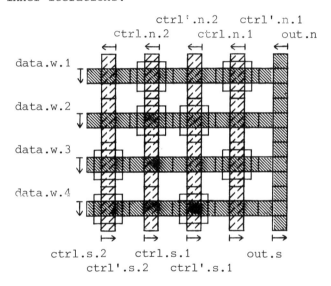

Fig.7 A selector.

It should be emphasised that the program in Fig.6 contains
no explicit geometric information, and this is to be expected
for many common VLSI subsystems. The double loop (array)
pattern is also very common in structural design, and many
other interesting examples can be produced by the use of
parameterization and recursion.

REFERENCES

Buchanan,I. (1980). "Modelling and Verification in Structured Integrated circuit Design", PhD thesis, Dept. of Computer Science, University of Edinburgh.

Burge,W.H. (1975). "Recursive Programming Techniques", Addison-Wesley, Reading, Mass.

Cardelli,L. (1981). "Sticks & Stones: An Applicative VLSI Design Language", Internal Report CSR-85-81, Dept. of Computer Science, University of Edinburgh.

Goguen,J.A., Thatcher,J.W. and Wagner,E.G. (1978) "An Initial Algebra Approach to the Specification, Correctness and Implementation of Abstract Data Types", in "Current Trends in Programming Methodology", Vol IV (ed. R.T.Yeh), Prentice-Hall.

Gordon,M.J., Milner,R. and Wadsworth,C.P. (1979). "Edinburgh LCF", Lecture Notes in Computer Science, N°78, Springer-Verlag.

Mead,C. and Conway,L. (1980). "Introduction to VLSI Systems", Addison-Wesley.

Milner,R. (1979). "Flowgraphs and Flow Algebras", Journal of the ACM 26(4).

Rowson,J.A. (1980). "Understanding Hierarchical Design", PhD thesis, California Institute of Technology, Dept. of Computer Science.

Williams,J.D. (1977). "Sticks; A New Approach to LSI Design", Master's thesis, Massachussets Institute of Technology, Dept. of Computer Science.

The DPL/Daedalus Design Environment

John Batali Neil Mayle Howard Shrobe
Gerald Sussman Daniel Weise
MIT Artificial Intelligence Laboratory
Cambridge, Massachusetts, USA 02139

The DPL/Daedalus design environment is an interactive VLSI design system implemented at the MIT Artificial Intelligence Laboratory. The system consists of several components: A layout language called DPL (for Design Procedure Language), an interactive graphics facility (Daedalus) and several special purpose design procedures for constructing complex artifacts such as PLAs and microprocessor data paths. Coordinating all of these is a generalized property list data base which contains both the data representing circuits and the procedures for constructing them. This paper will first review the nature of the data base. It will then turn to DPL and Daedalus, the two most common ways of entering information into the data base. The next two sections will review the specialized procedures for constructing PLAs and data paths; the final section will describe a tool for hierarchical node extraction.

1. The Data Base

The DPL system is organized around a hierarchical, object oriented data-base, written in LISP, which contains representations of designs.

The designer writes programs in DPL ("design procedures") that create and manipulate the data base. He may then query the data base to see the results of his work. Programs have been written in DPL that perform some of the work of design automatically.

One of the kinds of objects in the DPL data base is a TYPE. A TYPE contains a procedure that constructs a description of a design in the data base. These procedures take parameters, the parameters may take default values and constraints may be imposed among them.

A TYPE is essentially the procedural description of a "kind" of object, the parameters specify precisely what form the type is to take. For example, "enhancement-mode transistor" is a "kind" of layout structure, but many varieties may be made, they differ according to the dimensions of the channel region. In

---

This research was conducted at the Artificial Intelligence Laboratory of the Massachusetts Institute of Technology. Support for the Laboratory's VLSI research is provided in part by the Advanced Research Projects Agency of the Department of Defense under Office of Naval Research Contract Number N00014-80-C-0622 and in part by the Advanced Research Projects Agency under Office of Naval Research contract N00014-75-C-0643.
John Batali is an IBM graduate fellow.
Daniel Weise is a Fannie and John Hertz fellow.
Shrobe, Sussman and Mayle are the rest of the fellows.

DPL the enhancement-mode fet could be defined as a type, and the dimensions of the channel could be the parameters.

Complicated objects are described by creating type descriptions that "call" other types. The object created by this call will be "part" of the object created by the calling type. Structures thus created are called PROTOTYPEs in DPL. A PROTOTYPE of a type is a description of the structure created by calling the type with a particular set of parameters.

All prototypes of a given type, with a given set of parameters, are identical. To efficiently exploit this fact, instead of actually creating a new prototype every time a type is called with a set of parameters, DPL first determines, from information associated with the type, whether a prototype with the same parameters has been created. Only if none exists will a new prototype actually be created by running the procedure of the type. In either case, what is actually used in the description is not the prototype itself, but a VIRTUAL-COPY of it. A VIRTUAL-COPY is a "pointer" to a prototype. It "looks like" the prototype it is a copy of but the information is actually stored only on the prototype.

In many cases, the "appearance" of an object depends on its "context". For example, the location of a particular circuit object, viewed from the object it is immediately a part of, must be geometrically transformed when it is to be viewed from the chip as a whole. Also, the electrical description of a simple circuit will change, depending on whether it is placed, for example, in such a way that a wire shorts two of its nodes together. This contextual information is associated with a virtual-copy in a structure called an INSTANCE. An INSTANCE contains a VIRTUAL COPY and the context-sensitive information -- called its AUGMENTATION.

1.1 Summary of Data Base Structure

When parts are created in DPL it is possible to give the resultant instance a name. DPL contains a general mechanism for "extracting" information from the data-base by following a path of such names, automatically applying the context sensitive information as the information is brought out of the data base. Thus geometric points are transformed; and node descriptions are merged.

This *pathnaming* facility provides the designer with the ability to specify new structure of his design in terms of previously created structure. In this way the dependencies between objects allows the system to automatically determine what portions of the data-base must be reconstructed if parts of it change. Thus updating of the data base is automatic and allows for an *incremental* design style -- if the designer makes a small change to the design, the system can do most of the detail work needed to update the rest of the design.

The DPL data base can be used by programs manipulating several representations of a design. The "layout" and "electrical" representations have already been mentioned. In addition, the designer can use the data base in various ways to examine his design. Graphics capabilities exist in the Daedalus system for viewing the mask layers; CIF may be produced for the actual fabrication of designs; the electrical description of a design specified as artwork may be extracted.

2. The Language DPL

DPL is the name given to the set of LISP programs used to create and manipulate the data base. The most developed part of DPL deals with the layout representation of the design. This system has been used by relatively naive users in VLSI design courses at MIT to construct class project chips with good results.

The best way to present the features available to the user of DPL is to present an example. The following is the definition of a DPL type which will be named INVERTER. A picture of the result of calling this type with a certain set of values for its parameters is presented in Figure 1.

```
1   (DEFLAYOUT INVERTER
2       ((PRIMARY-PARAMETERS ((DL 8.0) (DW 2.0)
3                              (EL 2.0) (EW 2.0)
4                              PUZ PDZ Z))
5        (CONSTRAINTS ((C* DL PUZ DW)
6                      (C* EL PDZ EW)
7                      (C* PUZ Z PDZ))))
8       (PART 'PULLUP STANDARD-PULLUP
9             (CHANNEL-LENGTH (>> DL)) (CHANNEL-WIDTH (>> DW)))
10      (PART 'PULLDOWN RECT-E-FET
12            (CHANNEL-LENGTH (>> EL))
13            (CHANNEL-WIDTH (>> EW)))
14      (ALIGN (>> PULLDOWN)
15             (>> TOP-CENTER DRAIN-DIFFUSION PULLDOWN)
16             (>> DIFFUSION-CONNECTION PULLUP)))
```

This code describes the layout of a kind of object that will be called INVERTER. The inverter takes a number of parameters (lines 2-4). The parameters include DL, the length of the depletion channel, DW, its width, EL the length of the enhancement channel and its width, EW. The pullup ratio PUZ, and pulldown ratio PDZ, and the inverter ratio Z, may also be specified. The first four parameters take "default" values -- the default DL is 8, etc. Thus if the values for these parameters are not explicitly specified when the inverter is called, they will take these values.

The lines 4-7 of the code specify constraints among the parameters. These constraints simply describe the mathematical relationships between the parameters, for example the value of DL must be the product of the values of PUZ and DW. The constraint system of DPL is such that, only the values of enough parameters must be specified to constraint the rest according to the constraints declared. The system automatically uses the constraints to compute the values of the constrained parameters before the object is created.

The lines 8-13 create parts of the inverter by "calling" other, previously defined, types. Lines 8 and 9 call the type STANDARD-PULLUP and name the resulting instance PULLUP. The pullup is called with a certain set of values for its parameters, the values depend on the values of the parameters of inverter.

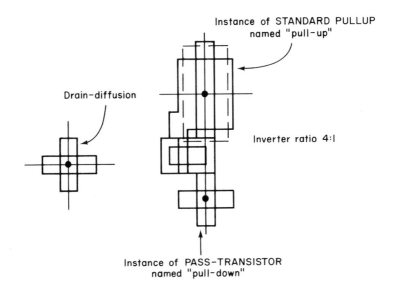

Fig. 1. An Inverter

`>>` is the name of the DPL accessing function. As used in line 9, it retrieves the values of the parameters of object in which it appears. `>>` can also be used to follow a path to extract other information from the data base.

Lines 10 through 13 create the pulldown for the inverter. As above, parameters are passed to the type being called by referring to parameters of the inverter.

In lines 14-16, the pulldown is moved to the correct location. This is done by specifying that a point on the pulldown must be aligned with a point on the pullup. The form on line 15 accesses the point which is the "top center of the part named 'drain-diffusion' of the pulldown". In the same way, line 16 accesses a point on the pullup.

Note that the pulldown of the inverter is placed by referring to another part of the inverter, rather than by explicit numerical coordinates. This ability for relative placement and specification of location is vital for an incremental design style. For example, in figure 2 we present a situation where a designer wishes to run a wire around an object. In the absence of relative positioning, he must specify the numerical coordinates of the point.

Suppose now, the designer decides that the object must be moved, and specifically it must be moved so far to the right that it will now inadvertently overlay the wire. He must now respecify the position of the wire, and indeed all parts of the design that are affected by the change. In general, this could be an arbitrarily large fraction of the design.

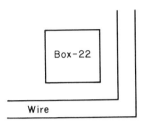

Fig. 2. An Example of Symbolic Relative Positioning

In DPL, the object could be named box-22, and the point could be specified as a certain offset from the lower-right corner of the object, for example:

```
(>> (pt-to-right (+ *poly-to-poly* (half *min-poly-size*)))
    (pt-below (+ *poly-to-poly* (half *min-poly-size*)))
    bottom-right box-22)
```

This specification is still valid even if the position of the object changes. Thus, merely by reexecuting the definition of this piece of the design, the data base may be reconstructed correctly. In addition to providing this ability for relative specification, DPL also automatically determines which portions of the data base must be reconstructed, and does so automatically.

3. The Daedalus Graphics System

Daedalus is an interactive, graphical interface to the DPL database implemented on The Lisp Machine, a personal computer designed and built at MIT. Daedalus uses a high resolution black and white monitor (850 by 1000 pixels), a color monitor (450 by 500 pixels), a "mouse" (a pointing device with three buttons) and an extended keyboard. Daedalus is now one of the main tools used for VLSI design at the MIT Artificial Intelligence Laboratory.

Daedalus may be thought of as an interactive, graphical programming environment for the DPL language. Its design embodies three central principles: *Transparency*, the *centrality of symbolic description*, and *incrementality*.

3.1 Transparency

The transparency principle asserts that "What you see is what you get." Two techniques have been used to meet this requirement. The first technique is that of active feedback: all of the basic commands of the system such as those which draw, copy, stretch and move boxes maintain a moving outline cursor which tracks the motion of the mouse; the command is completed by a second click of the mouse, producing the result shown by the cursor. Most of the commands are selected by pushing one of the 3 mouse buttons and some combination of 4 function keys (allowing 48 "immediate" commands). A special window on both screens tracks the selection of function keys, indicating what set of commands is available from the three mouse buttons. Most users have learned to use Daedalus quickly even in the absence of a system manual.

A second technique involves the use of a window system. VLSI designs are so large that a screen oriented system must choose between showing the details and showing the overall context. In contrast, Daedalus breaks each screen into several (possibly overlapping) graphic *views* each of which can be independently scaled and positioned. Views can be overlapped, buried and pulled to the top of the stack, allowing the user to quickly shift perspectives.

3.2 Symbolic Descriptions

Some parts of a VLSI design are best described by symbolic procedures (e.g. data paths and PLAs); others are more easily expressed graphically (e.g. the layout of a simple register cell). Unfortunately, there are a large number of objects in between these extremes. Many of the cells used in our layout procedures involve a significant number of parameters. The register cell used in the data path generator, for example, has a parameter which controls whether to extend one of the two signal busses; a second parameter controls the sizing of the power bus. Nevertheless, these parameters mainly specify variations off a single theme; it is preferable to describe the register cell graphically as long as the necessary parameterization can be captured.

In Daedalus, the user is able to express any information either symbolically (e.g. by typing an expression or DPL code) or graphically (by pointing with the mouse). Stored with each object in the current workspace is both the absolute coordinates describing the object's placement and parameterization and the symbolic forms (if any) which were used to describe the object. The absolute coordinates are used during display processing; the symbolic descriptions are used when creating DPL code. Similarly one may make changes either graphically or by editing the DPL code directly; in the second case the symbolic descriptions of the DPL code are copied onto the objects for use during further graphic editing.

In addition, during a graphic editing session, Daedalus maintains a set of parameters which parallel the parameters of a type definition. Three special mechanisms make use of this information. The first of these is a *fracture-line* mechanism which indicate stretch points in the cell; a parameter indicates the amount of stretching desired. The buffer can be stretched (as if it were a cell) by changing the value of this parameter. The second feature is a *conditionalization* feature. Any object in the buffer may have an associated condition (a symbolic form, involving the parameters); Daedalus has a display mode in which the object will appear if this form evaluates to *true*, otherwise it remains invisible. The third mechanism is a procedure which interactively develops a symbolic description for a particular parameter of an object, by walking through the path name hierarchy. This description can then be attached to an object. These capabilities made it possible to design many parameterized cells using only the graphics interface.

3.3 Incrementalism

The next design principle is that of incrementalism: small changes to the design should result in small amounts of work to achieve a new consistent state. Daedalus provides the ability to actively enforce constraints on the design space. Such constraints consists of a set of rules and a set of objects mentioned by the rules; each rule may mention several input objects and a single output object. A rule is triggered whenever one of its input objects is edited (moved, reparameterized, redefined, etc.); once triggered the rule is allowed to edit its output object to bring it into compliance with the intended constraint. Once the user tells Daedalus to impose the constraint all further processing is handled by the system, relieving the user of the need to make many tedious changes.

4. The PLA Generator

An important class of objects in the design of large digital systems is the Programmed Logic Array (PLA), actually, by the term PLA, we shall be referring to all very regular structures that perform some combinational or ROM type function. The regularity of Plas, they usually consist of several rectangular objects connected, with drivers, sense amplifiers, etc, make it possible to easily write programs that lay out a particular kind of PLA easily.

However, the broad usefulness of PLA-line objects suggests that an extremely useful tool would be a system that, rather than constructing a particular kind of PLA, supported generator programs and thus allowed the designer to create a PLA to his specifications. Such a system is under development. The system makes use of the data base of DPL to store information that the designer can use when laying out a PLA.

There are two important aspects to the design of a PLA. The first is the actual layout of the pieces. An ordinary PLA, for example, consists of a set of input drivers, an AND plane, an OR plane out output drivers. In some cases, however, the designer may wish to "fold" the PLA, of he may wish to add a selector to the output of the OR plane. In the PLA system, the designer has several planes -- input-drivers, and-planes, output-drivers, selectors. Each of these objects is a DPL type which contains auxiliary information that will be used when the planes are connected in ways specified by the designer.

Each plane contains information about how to connect to other planes. For example, the connection between a row of input drivers and an AND plane requires that the output signals from the drivers be connected to the signal lines of the plane. Also the planes, being types, contain parameters, the planes contain information about propagating parameter values during construction.

To construct a PLA the designer specifies the planes he wishes to use, and specifies the connectivity of the pla. The system constructs the planes, aligns, and connects them, and routes power and ground wires.

The second phase of PLA generation is "programming". The input and output of the PLA are abstractly defined in terms of "fields" which are collections of bits. A program for the pla is a list of pairings of field values on the input plane (which specify a product term) and the associated field values to be output.

This abstract program is then converted into actual programming of the previously created planes. Each plane has some effect on the values that pass through it. The driver for an AND plane, for example, produces from its input, two values, the input and its complement. This transformation is stored on the type of the object. The system uses this information to determine how the abstractly defined specifications are transformed to the actual inputs to the programmable parts of the PLA. This scheme also allows the PLA to be optimized by encoding certain subfields of the PLA and using a decoding PLA to produce the results. In this case, the decoding PLA accumulates its programming as it mediates between the coded field and the "real" output while the whole PLA is being programmed.

5. The Data Path Generator

The Data Path Generator is a tool for automatically constructing a microprocessor data path from a high level description of its contents. There are two components to this system and an organizing methodology.

5.1 Methodology

Although inspired by the Bristle Blocks [Johannsen] system, the Data Path Generator is intended to solve a quite different problem. Bristle Blocks is intended to lay out a general register machine; the Data Path generator lays out a special purpose register machine in which each register plays a unique role corresponding to a variable in the source microcode. Each register has associated with it a number of operators such as incrementers, constant sources, and testers. Connecting the objects in each such block is a *local bus*. Also running through each block is a *global bus*. The global bus continues from one block to the next, allowing inter-register communication. However, the local busses are independent; each register may have an operation performed on it during each cycle using the local bus as a communication channel. Thus, this architecture allows a high degree of concurrency.

5.2 Cell Construction

The data path is constructed as an irregular replication of cells each of which serves as a single bit slice of a register or operator. Each cell is therefore required to obey common methodological conventions. Six stretch points must be provided in each cell to allow correct sizing of power and ground busses and to allow stretching for pitch matching. In addition, certain types of cells may want to provide parameters for special common situations; these situations include special configurations of low and high order bits and bit-wise programmability (as in a constant source cell which may want to source either zero or one). These cells have all been designed via the Daedalus graphics editor three of whose features were added specifically for the Data Path Generator (these are fracture lines, conditionalization, and symbolic parameterization).

5.3 The Composer

The Data Path Generator per se is a composition program which selects cells from the library and places them in their appropriate places in the replication array. It also selects appropriate drivers and routes the connections between drivers and cells. The composer knows which registers and operators are bit-wise programmable and which are specially configured in the low and high-order bits, it uses this information to select the appropriate parameterization of each.

The greatest complexity faced by the composer is caused by the fact that the Data Path Generator is intended to support typed-pointer machines, thus each register may be broken into subfields (for type and datum) which need to be driven separately. When this occurs, the composer determines where to leave gaps for control and sense wire routing and stretches the appropriate cells to leave room for these gaps. The existence of subfields also means that registers and drivers need not pitch align; the composer program attempts to optimize the placement of drivers so as to minimize the routing between drivers and cells. Finally, the composer also tries to compact the empty space caused by registers or operators which only use one of the subfields.

6. Hierarchical Node Extraction

The schematic extractor of the DPL/Daedalus design environment fully exploits the structures and notions of DPL to operate hierarchically. It is also interfaced to the rest of the environment (notably Daedalus) and can be invoked by the designer at any time. The designer can then easily (*i.e.*, graphically or textually), inspect and use the results of the extraction.

The extractor operates hierarchically. It extracts each prototype only once and attaches the schematic to the prototype for further use by the designer or other analysis programs. In the sense that it exploits the hierarchy in this way, it is like the hierarchical analysis tools of [Whitney], [Hon], and [Scheffer]. However, the first one is concerned only with design rule checking, [Hon] is concerned with the general notion of doing all tasks hierarchically, and [Scheffer] places severe restrictions on the types of designs he can extract. [Hon]'s work is probably closest to ours, but none of the systems mentioned are as interactive or extensible as ours.

A schematic *virtually* consists of a listing of the the transistors and nodes of the prototype. The nodes and transistors of an object are retrieved using the standard accessor functions of DPL. Programs can be written which deduce electrical parameters from nodes and transistors as they record the geometric objects which define them. We say the schematic virtually exists because, just as the entire geometric layout is represented by geometric transformations upon its parts, the entire schematic is represented by "nodal" augmentations upon its parts.

This scheme extends the notion of an instance to include a NODE AUGMENTATION which states how the nodes of an instance map into the nodes of its parent. These augmentations are used to ensure that nodes retrieved from parts turn into nodes on their parent. For example, in the inverter example the source node of its pullup and the drain-node of its pulldown merge to form a new node. Retrieving either of the former nodes will return the latter node. This exactly parallels the notion of geometric transformations where points on instances are transformed to their parents coordinate system. One feature of this behavior is the ability to easily check connectivity of nodes --- if the nodes of parts map into the same node then they connect.

DPL has been extended to include commands for the naming, declaration, and verification of nodes. For example, one could include in the DEFLAYOUT for an inverter the forms:

```
(dnode 'output-node
       (>> source-node transistor)
       (>> drain-node pulldown))

(dnode 'input-node
       (>> gate-node pulldown))
```

When the prototype for the inverter is built the system checks that the gate-node of the pulldown is a node and then names that node INPUT-NODE on the prototype. It then checks that SOURCE-NODE of its transistor and DRAIN-NODE of its pulldown are nodes and that they intersect to form a new node which is then called OUTPUT-NODE.

The schematics can be output in a hierarchically based schematic description language developed by Chris Terman of the MIT Laboratory of Computer Science. Schematics described with this language can be simulated at the switch level using simulators Chris has developed.

References

[DPL] John Batali & Anne Hartheimer. "The Design Procedure Language Manual," *MIT AI Lab Memo 598*.

[Hon] Robert Hon. The hierarchical analysis of VLSI designs. CMU Computer Science Department VLSI Internal Memo Document V073. 1980

[Johannsen] Dave Johannsen "Bristle Blocks: A Silicon Compiler". *Proceedings of the 16th Design Automation Conference* pp. 310-313.

[Scheffer] Louis K. Scheffer. A methodology for improved verification of VLSI designs without loss of area. *Proceedings of the Second* Caltech Conference on Very Large Scale Integration), to appear.

[Whitney] Telle Whitney. A hierarchical design-rule checker, Masters thesis, California Institute of Technology, pending.

REGULAR PROGRAMMABLE CONTROL STRUCTURES

D.J. Kinniment

Department of Electrical and Electronic Engineering
University of Newcastle Upon Tyne, Newcastle NE1 7RU, UK

1. INTRODUCTION

One of the main aids to the control of complexity in the design of VLSI circuits is the provision and use of regular components with standard interfaces. To provide the maximum benefit these components must have sufficient generality within a fixed area to allow detailed changes of function without affecting other parts of the design. A good example of such a component, which has been widely used for implementation of finite-state machines in control applications is the Programmable Logic Array. The PLA allows a very efficient implementation of a synchronous state machine in NMOS technology and is easily programmed to provide a range of machine functions without affecting its logic structure.

There are, however, at leat two areas in the design of regular control hardware where the PLA in its present form is not ideal. Firstly there is a widely recognised need for the state machine to be implemented in a writeable medium. This occurs where, for performance reasons, it is important to add special new instruction to an existing machine. Such instructions are often application dependent and cannot be defined when the machine is designed. It is usual in such cases to use high speed RAM storing the microcode as part of a microprogrammed control system, but there are obvious area advantages in replacing this with a writeable PLA.

Parallel asynchronous control is a second area where it is important to provide regular structurs. There is an increasing pressure from the technology to make use of systems with as much local processing possible in order to reduce the necessity to move data long distances.

It is claimed that increasing clock distribution problems will also force the use of asynchronous systems to maintain timing independence and high performance, for these reasons it is necessary to examine how regular asynchronous control structures can be efficiently impelemented and where they can be applied.

Previous work by Patil (1975), has provided a basis for hardware implementation, but it is the purpose of this paper to extend that work and to provide comparisons in terms of NMOS technology for a number of techniques.

2. A WRITEABLE PLA

The PLA structure is generally thought of as an AND plane forming product terms, followed by an OR plane combining appropriate terms to give the required outputs. If this is to be made writeable, both AND and OR planes must have the programming links at each crosspoint replaced by a link controlled from a memory element. The memory elements themselves must also have a means of altering their information content according to some external data, and it is desirable to aid the testability of the system by providing a means to read the status of the stored information.

Performance optimisation depends mainly on optimising the memory cell area and read/write overheads since the interconnect structure of the PLA structure is already fixed.

2.1 The OR Plane Cell

It is not possible to replace each cell in the OR plane by a minimum area one transistor RAM cell since the destructive nature of the read out and the mechanism of the sense/refresh circuits make it impossible to select, read, and refresh two or more cells on the same data line with different data. The only available alternatives having non destructive read out are a 3 transistor dynamic cell or a static cell. The dynamic cell is shown in Figure 1, in which the charge on a storage capacitor controls whether the link T_2 is connected to ground via T_1 or not. T_3 allows a separate write/refresh route to the capacitor.

A static memory cell has the advantage of not requiring a separate refresh mechanism, but suffers from the disadvantage of large area and high power dissipation in NMOS technology.

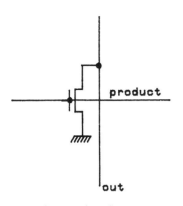 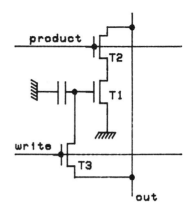

Fig. 1a) Fixed OR Cell   b) Dynamic OR Cell

2.2 The AND Plane Cell

Each link in the AND plane could be controlled in a similar way by either a 3 transistor dynamic memory cell or a static flip flop, but it should be noted that the number of stored states necessary to provide all possible product terms does not require a binary memory cell at each crosspoint in the matrix. An implementation of the AND plane could be provided by a single flip flop controlling the state of two links as shown in Fig. 2a.

This arrangement is similar to that normally used in a content adddressed memory (CAM): Aspinall et al (1968),so

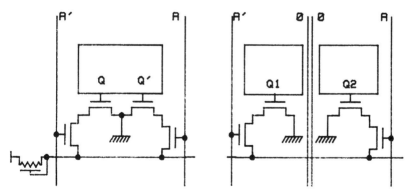

Fig. 2a) One Flip Flop Per Bit b) Two Flip Flops Per Bit
that the PLA as a whole would be implemented by a CAM for the AND plane followed by a linear select RAM the OR plane. Such systems have designed and built as address translation units, or cache memories in a computer system.

As shown the cell can select either the true or the inverse phase of the input variable A, but there is also a need to respond to neither in representing the 'don't care' case for that input. Such a need has already arisen in address translation units with variable size pages; Lavington et al (1971), and is dealt with by using two cells per bit as shown in Figure 2b. The same basic cell type can then be used for each case, but at a cost of doubling the hardware in the column containing the don't care case and <u>implementation</u> of a redundant state where both Q_1 and $\overline{Q}_2$ are set to 1.

Since three states are required, an implementation that only has three states should be more efficient, and can be provided by the tristate flip flop shown in Figure 3. Here only one of the three outputs. Q_A, Q_B, and Q_C can be low at any given time, thus in terms of the amount of hardware

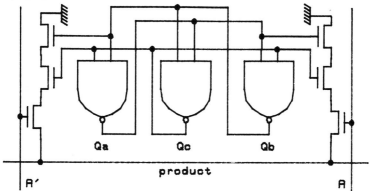

Fig. 3 Tristable Cell

and the power dissipation per cell a net gain has been acheived.

Even in the general case it can be shown that provided the area required for a three state cell is less than $\log_2 3 = 1.585$ times than that of a two state cell, a net gain has been made in the area required to store a given amount of information.

Whilst this advantage is either marginal or non existent in conventional information storage application the need for only three states in this case allows a three state cell with 1.5 times the area of a two state cell to occupy significantly less area than two two state cells.

In designing a practical array a means of writing in to the cell must be provided and testability is aided if conventional readout is also available. A simple method of acheiving both these is shown in Figure 4 in which the

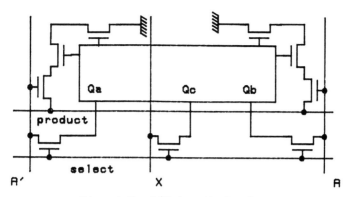

Fig. 4 Read/Write Mechanism

vertical lines are used both to change the state of the flip flops and to force one of the outputs to the low state.

2.4 Comparison of Techniques

Designs have been produced for both tristable and bistable static cells, and a 16x8 array of bistable cells has been fabricated for evaluation. These cells have the advantage of being useable in either the AND plane by using two cells per bit, or in the OR plane by rotating the cell through 90°.

The concept of a writeable PLA can only be evaluated by weighing its advantages in terms of flexibility against its disadvantages in terms of silicon area and power dissipation.

Here it is useful to compare each cell design with the conventional mask programmed PLA in each of these parameters.

TABLE 1
Writeable PLA Implementation

	cell area	Cells per bit AND	OR	Power Dissipation (active loads per bit) AND	OR
Mask Programmed	$64\lambda^2$	2	1	0	0
Dynamic	$\sim 500\lambda^2$	2	1	0	0
Static Bistable	$2419\lambda^2$	2	1	2	1
Static Tristable	$3840\lambda^2$	1	–	1	–

Table 1 shows that a minimum area dynamic writeable PLA would occupy approximately 8 times the area of a typical mask programmed design and that for an array in which the AND plane was implemented by tristable cells and the OR plane by bistables, about 33 times the area would be required.

3. PARALLEL ASYNCHRONOUS ARRAYS

Complex high performance machines in which the control is entirely self timed have been designed for many years. The control of such machines is often implemented by hardware which has the structure of a parallel flow chart and operates in a way analogous to a Petri net representation. Tokens are represented by signal changes and places by storage elements. The analogy is not exact and the design philosophy is generally not rigorous, with the consequence that the control structures are irregular and difficult to design.

3.1 Petri Net Matrix Representation

There is a body of theory regarding the operation of Petri nets and other directed graphs which makes the implementation of control structures based on such rigorous concepts attractive, and hardware realisation of Petri nets have been presented by Patil (1975). Patil observed that any network of places and transitions in which a place must always be followed by a transition and vice-versa could be arranged as a matrix with one column for each place and one row for each transition. The cross points of the matrix are of five types, place-transition, transition-place, null and two representing the value of a boolean variable controlling a transition.

A Petri net and its corresponding matrix realisation are shown in Figures 5 and 6.

In these diagrams the output level of the place circuit

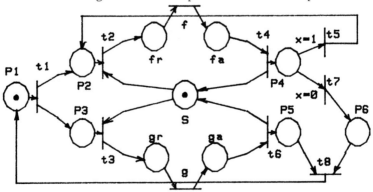

Fig. 5 Petri Net

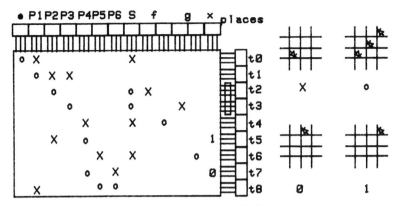

Fig. 6 Matrix Representation

is high when it is occupied by a token, thus transitions fire when all their input places, determined by an AND gate diode on wire A, are full.

When a transition fires all output places are filled by a token represented as a signal change from high to low on wire 1, and the input places emptied. The completion of this action is signalled by another AND gate on wire B which restores the transistion to its original state.

Since diodes can only exist on the crosspoints C_1, B_2, A_2 and A_3 it is possible to separate the four possible crosspoints locations onto two planes, A_2, A_3 and B_2 being part of an AND plane, and C_1 being part of an OR plane, as shown in Figure 7.

The two planes shown here have an almost identical function to the planes in a PLA, and could be implemented in the same way. The structure produced, however, is capable of parallel asynchronous operation with the place circuits holding the state variables in a less compact form than the conventional PLA state machine.

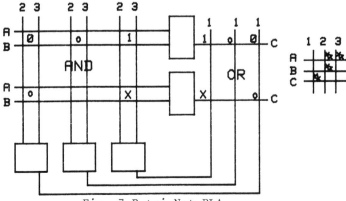

Fig. 7 Petri Net PLA

3.2 Arbitration

The use of self timed systems does not eliminate the metastability problems encountered with synchronisation, in fact the asynchronous nature of such systems can give rise to much more frequent occurences of timing clashes arising from arbitration between two asynchronous requests for the same resource. This situation typically occurs in a self timed computer where requests for data from the main or cache memory originate from two or more different and variable timed sources- e.g. the instruction prefetch unit and the floating point unit. Since each request is self timed, the timing of the two are, by definition, unknown and variable. They may therefore occur simultaneously, and frequently do.

This situation is dealt with in Patel's realisation by an input place shared between two transitions, thus only one transition may fire, absorb the token and prevent the other firing. Since the circuit realisation of such an operation involves the resolution of an infinitely variable quantity (time) into a binary value (which transition is to fire first) a special arbiter circuit is placed between the matrix outputs and the transition involved which ensures that only one transistion may fire. The location of this arbiter is shown in Figure 8, and can be see that the regularity of the array is detrimentally affected because the location of arbiters depends on the information stored in the matrix.

The design of arbiters is relatively straight forward; Kinniment and Woods,(1976) and will not be discussed in detail here except to present the structure of a general arbiter in Figure 9.

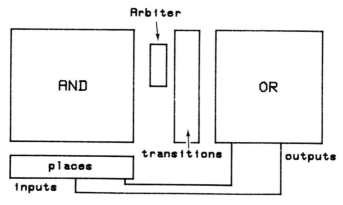

Fig. 8 Arbitration Location

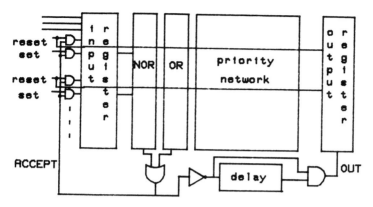

Fig. 9 Arbiter System

The input register of this arbiter consists of a single bit for each source of input requests, plus one bit for each extra signal used to control dynamic priority allocation. When any request bit is set by an external signal, a NOR gate connected to the request bit outputs responds by going low, allowing the ACCEPT signal to go low and preventing any further change in the input vector. A purely combinational network now evaluates which of the output register bits should be set, and after a delay corresponding to the decision time of the input register flip flops the output register is set by the signal OUT. Any reset signal resulting from the allocation of the resource causes ACCEPT to go high, corresponding bits in the input and output regsiter to go low, and if an outstanding request exists in the input register, the arbitration cycle to recommence.

Since the OR gates and priority network are purely combinational it is a simple matter to implement both of these as a PLA so that the overall control structure is

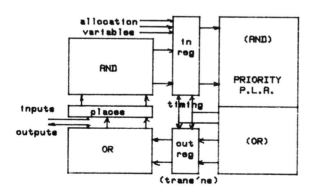

Fig. 10 Asynchronous Control

arranged as in Fig. 10.

This arrangement is approximately rectangular and provides more flexibility in its structure than the original.

4. CONCLUSION

Techniques have been presented which enable the implementation of writeable PLAs in NMOS technology and regular structures for parallel asynchronous control systems. The two could, of course, be combined to provide a completely general writeable asynchronous control. Whilst the writeable version of the PLA offers considerable flexibility, it occupies an area approximately 8 or 33 times greater than the mask programmable version. This disadvantage is likely to confine its application to a few areas, such as the provision of a few extra uncommitted functions in a computer system where most of the functions are fixed. The disadvantages of the asynchronous control are likely to be its relatively large size and slow operation which arise because of the relatively poor packing of instantaneous control state in the place circuits and the need for arbitration. A more compact coding of all the possible control states may be possible, but is beyond the scope of this paper. These and other problems likely to restrict the use of asynchronous control to applications where the time and area overheads are a small proportion of the system total, ie.control at the processor-memory-switch level or above.

5. ACKNOWLEDGEMENTS

The author would like to thank the SERC for the provision of facilities which made this work possible as well as Mr. McLauchlan of the Computing Laboratory, Newcastle University who gave considerable assistance with the design.

6. REFERENCES.

Aspinall, D., Kinniment, D.J. and Edwards, D.B.G.E. (1968) Associative Memories in Large Computer Systems, Information Processing, Vol 2, Ed AJH Morell, North Holland Publishing Co., pp 796-800.
Kinniment, D.J., and Woods, J.V.,(1976) Synchronisation and Arbitration Circuits in Digital Systems, Proc. IEE, Vol 123, No. 10 pp 961-966.
Lavington, S.H., Kinniment,D.J., and Knowles,A.E.(1971). An Experimental Paging System, Computer Journal, Vol 14, No. 1 pp 55-60.
Patil, S.S.(1975) Micro-Control for Parallel Asynchronous Computer, MIT Computation Structures Group Memo 120.

SESSION 5

VLSI CHIP DESIGN AT THE CROSSROADS

J. Craig Mudge

*VLSI Program, Division of Computing Research, CSIRO
Adelaide, S.A. 5063, Australia*

1. INTRODUCTION

As we enter the VLSI generation of computer technology, the fabrication of 100,000 transistors on a single silicon chip is possible. Moreover, the steady increase of transistors per chip shows no signs of slowing. Because of a few significant events of the last couple of years, the field of VLSI research is at a crossroads. This paper suggests some of the paths that will be chosen.
Some significant events are:
- the establishment of inter-disciplinary VLSI research programs at many universities.
- the demonstration, via Multi-university Project Chip (MPC) implementation systems, that a clear interface is possible between design and fabrication.
- moves by systems manufacturers to establish facilities for silicon design and fabrication

This paper discusses three topics:
1) formal composition systems;
2) design representations;
3) interface between design and fabrication;
in which there will have to be significant progress if we are to harness the technology available to us.

2. FORMAL COMPOSITION SYSTEMS

A composition system is a design aid for combining the constituent parts of a design to form an entire, correct chip. Figure 1 introduces the terms leaf cell (20 to 100 transistors) and composition cell. Triangles represent compositions; they are either composition cells or composition rules. Two cells are joined, by either a routing block (an example of a composition cell) or by abutment (a composition rule). By this convention, composition cells add no extra functionality to the design. The more obvious (because it is physical) form of composition joins cells containing layout data. For tractability, such cells are usually rectangular, although this is not essential. Composing other types of data, such as logic representations, will be discussed later.

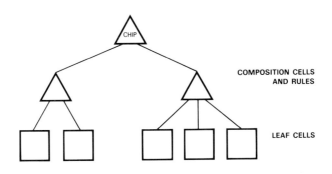

Fig. 1. *A hierarchy of composition cells and leaf cells*

The dual of composition is known as decomposition. In this discussion we assume that decomposition is done manually (at the earliest stage in a design) to form parts of tractable size. It is composition that we wish to formalise.

Composition is the very essence of complexity management and yet very few examples of such tools exist in 1981. On the other hand, computer-aided design tools abound for leaf-cell design. Hence, composition systems demands our attention as an important area of research.

Multi-Project Chips have stimulated innovation in leaf-cell tools at universities. Communications networks associated with such chip implementation systems further stimulate leaf-cell work. Suppliers of commercial IC design sytems have contributed as well; however, their systems are not keeping pace with device densities as we shall see later.

The few examples of true composition systems which exist in 1981 are either automatic-place-and-route tools or silicon compilers.

Automatic-place-and-route systems, employing gate arrays (Chen *et al.*, (1977), for example) and standard cells (Persky *et al.*, (1976), for example) compose leaf cells of logic representations. They have been used successfully by computer manufacturers, are now well understood, and

commercially available services are now offered.
Silicon compilers, on the other hand, are still in their
infancy, but demonstrations, for example, Bristle Blocks
(Johannsen, 1979) are promising. Both these classes are
outside of the scope of this paper and will not be discussed further.

DEC's FPP Experiment

In 1979, Digital Equipment Corporation in Massachusetts
formed a small group to develop a design methodology for
VLSI circuits. The group set out to combine recent research
results with the constraints and opportunities which obtain
in an industrial environment. The goals and results are
described elsewhere (Mudge *et al.*, (1980a, b) and Fairbairn (1981)). A key result was the demonstration of a
factor of 10 decrease in design time for VLSI circuits.

A structured design methodology and a chip-assembler tool
(CHAS) to support the methodology were developed during the
design of a 77,000 transistor floating-point-processor chip.
Looking back at the experience, it appears that a true composition system was created.

CHAS is a data manager for the manual (ad hoc) composition
of three representations (layout, circuit, and logic).
Below we highlight four key results of this experience as
they pertain to composition systems.

1. Floor Plan as Gateway Figure 2 shows the key role of a
chip floor plan in the design methodology. A designer
operates on his logic, circuit and layout data with the aid
of a command language in CHAS. All accesses to a chip data
base, whether they are for creation, analysis or composition
(assembly) are via names of rectangles in the floor plan.
Several consequences are apparent: data-base integrity is
aided, since a change to one representation causes others to
be marked as suspect; global optimisations are more likely
to occur; and geometric, ie topological, considerations are
more likely to dominate logical considerations in the design.

2. Rapid Changes At about two-thirds of the way through
the design of this chip, we had to make a major change to
the floor plan to satisfy die-cavity constraints. The die
dimensions were approximately 10.2mm by 5.1mm and needed to
be more square. Figure 3(a) shows the rectangular floor
plan, which was changed to one closely resembling the final
floor plan seen in Figure 3(b) which has a die size of
7.4mm by 7.4mm. The major perturbations can be seen in the
figure: the PLA group was placed between the fraction and
exponent data paths, and the ROM moved from the right to
the top left of the chip.

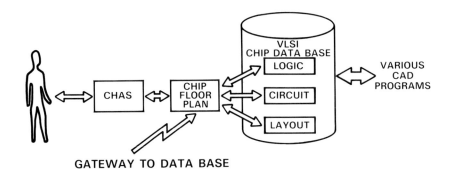

Fig. 2 With CHAS, a prototype chip assembler, a designer operates on his logic, circuit and layout data by accesses through a chip floor plan

The effort to reassemble (recompose) the chip in the second form was about eight hours of a technician's time at our experimental graphics work station. I estimate that if this error-prone process were to be done manually, three months would have been required.

3. *Hierarchy of CAD Programs* A commercial IC design system, a CALMA GDS-1, was a part of our tool set. As with other contemporary systems, it was not possible to map its model of chip design on to our design methodology. However, we used it extensively to prepare layout data for leaf cells; such data were converted to CHAS format and stored in a chip data base. A designer then used CHAS to compose these leaf cells with cells created by other tools.

The relevant observation is that the commercial design system was a component of our overall design system, and hence subordinate in the hierarchy of tools. This contrasted with the conventional use of such commercial systems.

4. *Project Management* The floor plan was an effective tool for keeping team members informed of design status.

The Future

The examples given above point to the need for research in *formal* composition systems. Rowson's work (1980) at California Institute of Technology is an excellent start. He examines hierarchical equivalence, an important problem in composition systems, as we shall now see.

As designers we take several different views of a chip as we proceed with a design. These different views employ different representations; some examples (in addition to layout, circuit and logic mentioned previously) are:

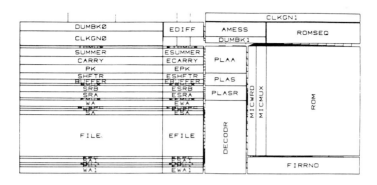

(a)

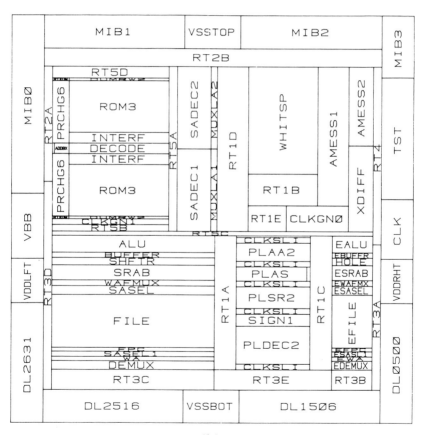

(b)

Fig. 3 The chip floor plan of a 77,000 transistor floating-point processor at two stages in its design

symbolic, register-transfer, processor-memory-switch, sequence, delay, and power. The problem of composition becomes much worse when we are composing multiple representations. Composition rules may differ for different representations as shown in Figure 4; layout blocks are additive, and yet delay is additive when joined serially and not changed (the identity function) when parallel.

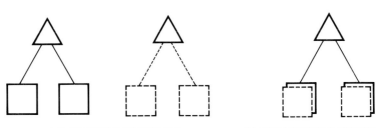

(a) TWO DIFFERENT COMPOSITION FUNCTIONS   (b) SAME COMPOSITION FUNCTION

Fig. 4 When more than one design representation is used, the composition functions applied in constructing a hierarchy may not be the same, and yet we need to prove hierarchical equivalence

A composition data structure will be invented. Such a structure will contain the necessary information from a leaf cell description such that when one or more leaf cells are composed, statements can be made about the combined entity (its correctness, for example). For the layout portion of such a data structure, we know that input-output connections need to be enumerated, but for delay and power, what should be brought to the edges so that a composition algebra can be formulated?

Methodology

A final observation is that in each of the successful examples of VLSI chip design, it appears that the design methodology was more important than the particular suite of CAD tools.

3. NEEDED: A SPANNING REPRESENTATION

Carver Mead stated our major problem when he wrote (Mead, 1979):

"VLSI is a statement about system complexity, not about transistor size or circuit performance. VLSI defines a technology capable of creating systems so complicated that coping with the raw complexity overwhelms all other difficulties".

As we try to cope with complexity, we are also trying to
a. approach optimum use of the silicon medium;
b. devise team organisations that are effective for multi-person designs, and
c. deal with the scarcity of designers.

The previous section of the paper discussed different representations in the context of composition systems. This section raises some questions about how the different representations interact.

Now that we can put complete computer structures on a single chip, we find that disciplines which were once separated by natural, clean interfaces (circuit design on a chip and logic design on a printed-circuit board) are now coexisting.

How do we marry these disciplines?

As we look at the disciplines individually, we see that some of them, as traditionally practised, are in conflict.

Conflicting Skills

The layout artist, trained to pack circuits into the smallest possible area, produces exquisite patterns which, when combined with other sections, produce a chip whose global interconnect is usually quite random. As a result, overall chip speed and area suffer.

A logic designer, trained to minimise the number of logic elements, such as gates, measures himself by the degree to which he can minimise transistor count. Of course, wiring suffers again. Two factors reinforce this: his most formal tool is logic minimisation and his representation (a logic schematic of connected gates) reflects little of the physical properties (power and ground wiring, in particular) of the silicon medium.

A circuit designer, in optimising speed and power, produces circuits with a variety of width-to-length ratios, and so overall chip regularity suffers.

So we see that there is an inherent conflict between the three.

Division of Labour

Horizontal and vertical divisions are possible ways of partitioning a multi-person design. The horizontal division parcels out separate sections to people of the same skills; the vertical division does not subdivide the chip, but stratifies it into different levels of representations.

An example, *par excellence*, of the vertical approach is the design methodology used to develop three Intel iAPX-432 chips (Lattin *et al.*, 1981). The methodology identifies a set of design levels and a rigid (manual) translation between the levels, which are
- architectural design
- logic design
- circuit design
- layout design

The distinctions are strictly maintained and computer programs at each level process descriptions of the design at that level. Special care was taken to ensure bilingual abilities of the design staff so that a designer could understand the languages used at the next higher level.

There is no doubt that the methodology worked, and yet one is led to suspect the optimality of the result, given our earlier discussion of the inherent conflicts between the representations.

DEC's design team for the FPP chip experiment also used a vertical division of labour. Perhaps the floor plan provided a greater span (range of vision) across the representations.

For the horizontal approach, one needs designers who can span the disciplines involved. Most MPC designs have been done by such designers. However, the chips so designed are not of VLSI complexity, nor do they employ the type of local optimisations in circuit and layout sometimes necessary for products.

A Comparison with Programming

Programming is a more mature discipline than chip design in that programmers faced complexity management about 10 to 15 years earlier than chip designers. Table 1 compares the design representations used in the two disciplines as they are practised today.

TABLE 1

An Analogy with Programming Suggests that there is a Superfluous Level in Chip Design

DOCUMENTATION	PROGRAM	CHIP
One-page view	structure graph	floor plan
Intermediate representation	algorithm description	logic diagram circuit diagram
Executable design	source code	mask layout

Programming went through a stage of having programming and coding as distinct skills. As higher-level languages developed the distinction disappeared and no longer did a programmer deliver blow-by-blow charts to a coder.
With programming history as a guide, I expect chip design to go through a stage of shedding a superfluous representation.

4. INTERFACE BETWEEN DESIGN AND FABRICATION

The idea of a clean interface between design and fabrication (Mead, 1979) is now a reality. The success of MPC implementation (Conway et al., 1980) at various U.S. universities has shown that it is indeed possible to design at a site remote from one's wafer fabrication capability. We are now witnessing the next stage, that of going from prototype demonstration to full-scale concept realisation.
As the silicon foundry comes to reality and more systems manufacturers acquire their own captive semiconductor facilities, what can we expect?

Experimental Computer Architecture

Since early in the second (transistor) generation of computer technology, most advances in computer architecture have occurred in the laboratories of computer manufacturers. Because the art has been technology driven, and because the rate of change in technology has been so rapid, very few universities have built influential machines. (However, there are notable exceptions, for example, Manchester University and Carnegie-Mellon University.)

In the 1980s, however, it will be economically feasible for university researchers to build machines again - complete systems on silicon. Thus experimental computer architecture will revert to its birthplace during the first (valve) generation.

Standard Processes

Historically, standard processes (such as those used for MPC designs) have had more relaxed design rules than leading edge production processes. This relaxation can mean as much as a two-generation difference (factor of four) in performance.

Designers of systems on silicon will reach the point where, in order to produce competitive products, they can no longer tolerate such penalties. Silicon foundries will respond by radically improving the characterisation of their advanced processes.

Furthermore, there is likely to be a divergence between processes for standard high-volume chips and processes for custom chips. Although equivalent in minimum feature size, a process for custom chips will be oriented towards the special needs of the system designer. We can already see this happening in the case of interconnect layers in MOS technology. A second layer of metal has been shown to shorten design time. Hence, process developers with custom chips in mind have diverged from the traditional evolution path followed by the producer of standard chips, who finds a second layer of polysilicon to be adequate.

International Design Centres

Already we have seen the establishment, in foreign countries, of design centres by two major U.S. semiconductor firms. These moves have been motivated by the scarcity of chip designers. As more silicon foundries emerge, this trend will accelerate.

References

Chen, K.A., Feuer, M., Khokhani, K.H., Nan, N., and Schmidt, S. (1977). The Chip Layout Problem: An Automatic Wiring Procedure. *Proceedings of the 14th Design Automation Conference*, 1977

Conway, L., Bell, A., and Newell, M.E. (1980). MPC 79: The Large-Scale Demonstration of a New Way to Create Systems in Silicon. *LAMBDA 1*, 2, 10-19

Fairbairn, D.G., (1981). VLSI Design Tool Development at DEC. *LAMBDA*, 2, 1, 12-13

Johannsen, D. (1979). Bristle Blocks: A Silicon Compiler. *Proceedings of Caltech Conference on VLSI*, January 1979, 303-310

Lattin, W.W., Bayliss, J.A., Budde, D.L., Rattner, J.R., and Richardson, W.S. (1981). A Methodology for VLSI Chip Design. *LAMBDA*, 2, 2, 34-44

Mead, C.A. (1979). VLSI and Technological Innovation. *Proceedings of Caltech Conference on VLSI*, January 1979, 15-28

Mudge, J.C., Peters, C., and Tarolli, G.M. (1980a). A VLSI Chip Assembler. *In* Design Methodologies for Very Large Scale Integrated Circuits, NATO Advanced Summer Institute, Belgium, 1980

Mudge, J.C., Herrick, W.V., and Walker, H. (1980b). A Single-Chip Floating-Point Processor: A Case Study in Structured Design. *In* Design Methodologies for Very Large Scale Integrated Circuits, NATO Advanced Summer Institute, Belgium, 1980

Persky, G., Deutsch, D.N., and Sweikert, D.G. (1976). LTX: A System for the Directed Automatic Design of LSI Circuits. *Proceedings of the 13th Design Automation Conference*, 1976

Rowson, J.A., (1980). Understanding Hierarchical Design. Ph.D. Thesis, California Institute of Technology, 1980

A HIERARCHICAL DESIGN ANALYSIS FRONT END

Telle Whitney

Department of Computer Science
California Institute of Technology
Pasadena, California

INTRODUCTION

This paper presents a design style aimed at reducing the complexity of designing a VLSI circuit, and then presents an algorithm which exploits this design style in order to reduce the computational complexity of analyzing a large design. Hierarchical design is a design methodology which allows a designer to break a design into smaller, more manageable pieces. This design style will be presented along with an algorithm which exploits the hierarchy to allow an analysis of very complex circuits. An example implementation of this algorithm for geometric design rule verification is then described in detail.

HIERARCHICAL DESIGN

A hierarchical design is a design which is defined as a composition of other smaller designs. A hierarchical design style is based on a methodology that assumes that a problem can be broken down into a set of smaller problems. This division of the problem continues until a simple solution for each of the small problems is feasible. Then the primitive solutions are combined together to form the larger solution. This approach to problem solving is a hierarchical approach (Simon, 1962).

The following terms will be used throughout this paper. A symbol is a collection of elements. An element is either an instance of a symbol or a component. A component is either a piece of geometry or a structural element. A piece of geometry is a wire, a box or a polygon. Example structural elements are transistors, or contact cuts. An instance of a symbol is a reference to a symbol with placement and orientation information. A bounding box is a rectangle which bounds all of the element's internal elements. The Minimum Bounding Box (MBB) represents all elements during the composition process.

A design system which supports hierarchical design allows a designer to do three things: 1) create symbols, 2) compose two symbols, and 3) compose components and symbols. A designer creates a symbol by placing many elements together in some order. A designer composes two symbols by placing these symbols together according to some rule. A designer composes a component with a symbol by placing these two elements together. The process of creating a symbol is nothing more than the repeated application of composing two elements.

The quality of the instantiation of the hierarchical design is determined by the composition rules used. So it is important that the composition rules are chosen carefully. Jim Rowson (1980) in his PhD thesis proposes a set of very strict composition rules for a design system. In this proposed system the MBB of two symbols cannot overlap. Also, symbols cannot contain both symbol instances and components. This simplies the automatic process of composition and guarantees design correctness but results in overly conservative designs. A set of null composition rules is at the other end of the spectrum. A design system using null composition rules would allow elements to be placed together in an arbitrary manner. This allows the designer to have total control over the design, but makes it difficult to generate correctness. A usable set of composition rules are probably somewhere between these two extremes.

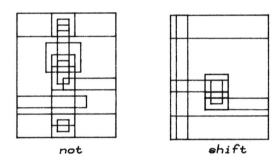

Figure 1 - Two Symbols

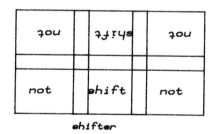

Figure 2 - Composition Symbol

Figures 1 through 3 illustrates the results of a possible set of composition rules. Symbols' and components' MBBs are allowed to overlap enough to ensure connectivity. Figure 1 shows two symbols: 1) an nmos inverter called "not", and 2) a pass transistor called "shift". Figure 2 illustrates a symbol called "shifter", which is composed of several instances of the symbols "not" and "shift". The MBBs of the symbol instances overlap to ensure connectivity between the two instance's signals. The symbol instances in the second row are mirrored in order for the two rows to share a power bus. Figure 3 illustrates a symbol called "dbl_shift" which is one more level of composition in the hierarchy. This symbol contains the connection wires to the power and ground buses in the internal symbol instances of "shifter". This example illustrates a possible set of composition rules whose use would define a particular type of hierarchy having the properties of relatively easy automatic generation, while ensuring correctness.

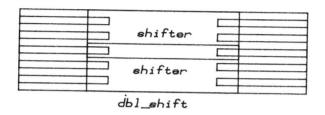

dbl_shift

Figure 3 - Second Composition Symbol

Hierarchical design encourages a designer to introduce regularity into a design. When a designer finds a solution to a problem, the hierarchy allows this same solution to be used in many different contexts. If a designer uses this approach, then there is clearly a great deal of redundancy in the design.

A HIERARCHICAL FRONT END ALGORITHM

Given designs defined in a hierarchical manner, it is possible for design analysis tools to exploit this structure. Using this strategy, it is possible to significantly reduce the amount of information a programs must process. This section introduces an algorithm which may be used as the front end for many different design analysis tools.

The algorithm must examine two types of relationships in order to determine functional or geometrical violations. First, the algorithm must look at each symbol definition. Second, it must examine each interaction between two elements. Both of these situations need to be looked at only once, the first time they are encountered in the design, and not each time they are present in the description. The algorithm does this by recording two things: 1) symbol definitions previously checked, and 2) interactions previously checked.

The data structure the algorithm uses is a set of symbol definitions. One of these definitions is the entire design, all others are part of the design. Each symbol definition has four pieces of information: 1) a bounding box, 2) a list of elements, 3) an interaction list, and 4) a checked flag. The list of elements contains all the elements which the symbol uses. The interaction list contains the information about the interaction checks involving this symbol which have been previously performed. The checked flag indicates whether this definition has been checked.

An element may be either a component or a symbol instance. A symbol instance contains three pieces of information: 1) a bounding box, 2) a reference to the symbol definition, and 3) a transformation describing the placement of the symbol. A component contains two pieces of information, which are: 1) a bounding box, and 2) the position information. The position information describes the placement of the component.

The outline of the algorithm for processing a hierarchical design is shown in figures 4 through 6. The algorithm performs the same process as the design system did only in reverse. Just as hierarchical design implies breaking the problem up into smaller problems and then composing these solutions together, the hierarchical algorithm breaks the design apart until it can perform some primitive operations, and then looks at each unique interaction defined by a composition.

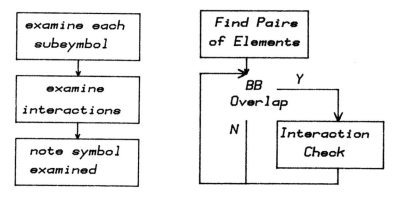

Figure 4 - The Algorithm   Figure 5 - Examine Interactions

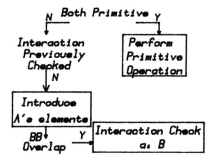

Figure 6 - Interaction Check A B

Figure 4 illustrates how the algorithm processes a symbol. A symbol is processed by examining the symbols referenced by the current symbol, and then examining each unique interaction between the symbol's elements. The complete design is the initial symbol the algorithm processes.

Figure 5 illustrates how the algorithm examines the interactions of a symbol's elements. The algorithm finds each pair of elements which interact and then looks at that interaction in detail. A pair of elements interact when their bounding boxes overlap. If the description of the design is a geometrical description, then the elements' bounding boxes must be bloated before interactions are determined. The factor by which the bounding boxes are bloated is determined by the largest applicable spacing rule of the technology in which the design is to be implemented. If the design description is a circuit or gate level description, the boxes must also be bloated according to the description's spacing between wires and components.

Figure 6 illustrates how the algorithm examines a particular interaction between elements A and B. If the two elements are both components, then no further breakdown is possible. At this point primitive operations are performed. These operations are determined by what type of analysis is required for the particular problem. (Some examples of primitive operations for geometrical design rule checking are presented in the next section.) Otherwise, the algorithm determines whether or not this particular interaction has been previously examined. It does this by looking up the interaction in the interaction list associated with one of the symbols. If both elements are symbols then the choice of which interaction list to look at is arbitrary, since the interaction is stored with both definitions. If the interaction is between a component and a symbol, then the interaction is stored with the symbol.

If the interaction has not been previously examined, the algorithm must now look at the interaction. The algorithm does not have enough information at this point to determine anything, so it must introduce new information. It does this by using the elements of one of the symbols, and performing further interaction checks. If only one of the elements is a symbol, then it is obvious which symbol to break apart. If both elements are symbols, then the choice of which symbol to break apart is arbitrary, and may be optimized for a particular implementation.

This algorithm only works if the primitive operations for a particular type of analysis can be defined in terms of a set of comparisons between two elements. This is possible in geometric applications if all structural elements, e.g. transistors and contact cuts, are explicitly present in the description. The primitive operations for design rule checking, will be described in the next section.

A DESIGN RULE FILTER IMPLEMENTATION

A hierarchical design rule filter at Caltech implements the algorithm described in the previous section (Whitney, 1981a,b) for geometric design rule analysis. The filter takes a design description in the Caltech Intermediate Form (CIF) (Mead, 1980) and discards elements that are redundant. The program produces a fully instantiated geometry file which can then be used as input to a traditional geometric Design Rule Checker (DRC) program. The program contains little specification of geometrical properties and uses a minimal specification of the design rules associated with the implementation technology.

The filter restricts the structure of the CIF file in order to maintain the integrity of the algorithm at all levels of the hierarchy of the design. A call to the top level symbol is the only command allowed outside of a definition. This final call is a reference to the top level symbol definition and therefore does not have a transformation associated with it. Checking the design is equivalent to checking a symbol definition with this type of structure enforced. Symbols not referenced are not checked.

The filter is implemented both in SIMULA on a DEC 2060 and in Pascal on a VAX 11-780. The following algol-like description describes the flow of the program:

```
!Top Level Definition;
BEGIN
    scan_cif_noting_all_symbol_definitions;
    examine_symbol(top_level_symbol);
END;


FUNCTION examine_symbol(symbol);
BEGIN
    FOR each symbol_element DO BEGIN
        IF symbol(symbol_element) THEN BEGIN
            IF NOT examined
                THEN examine_symbol(symbol_element);
                note_bounding_box;
            END ELSE note_bounding_box;
    END;
        examine_interactions_between_symbol_elements(symbol);
        mark_symbol_examined;
    END;
END;


FUNCTION examine_interactions_between_symbol_elements(symbol);
BEGIN
    sort_elements_in_x_and_y;
    FOR each a,b which_may_interact DO BEGIN
        IF
        bloated_bounding_box_overlap(symbol_element(a), symbol_element(b))
        THEN BEGIN
            IF (symbol(symbol_element(a)) OR symbol(symbol_element(b)) THEN
                examine_element_interactions
                (symbol_element(a),symbol_element(b))

            ELSE
                design_rule_examine(symbol_element(a),symbol_element(b));
            END;
        END;
END;

FUNCTION examine_element_interactions(element1, element2);
BEGIN
```

```
IF symbol(element1) OR symbol(element2) THEN BEGIN
    element1:-which_element_is_symbol;
    elemnt2:-other_element;
    FOR each element1_element DO
        IF bloated_bounding_box_overlap(element1_element, element2)
            THEN examine_element_interactions(element1_element, element2)
    END ELSE design_rule_examine(element1, element2);
END;

FUNCTION design_rule_examine(geometry1, geometry2);
BEGIN
    IF design_rule_exists(geometry1, geometry2) THEN BEGIN
        write_out_cif(geometry1);
        write_out_cif(geometry2);
    END;
END;
```

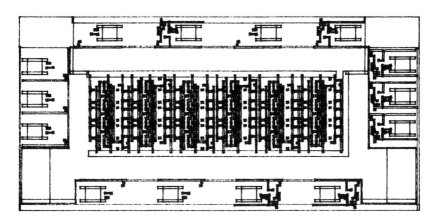

Figure 7 - FIFO Initial Geometry

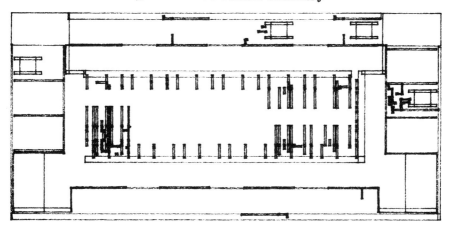

Figure 8 - FIFO Filtered Geometry

The filter uses a simple definition of the primitive operation. If a design rule exists between two pieces of geometry, then write the geometry out to a file. When the filter has finished, this file will contain all possible design rule violations. This file of fully instantiated geometry can then be used as input to a traditional geometric design rule checker. This design rule checker will then be able to determine all real design rule violations.

Figure 7 is a plot of a FIFO, designed by Eric Barton for the VLSI course at Caltech. Figure 8 is a plot of the remaining geometry, after the design was processed by the filter. This filtered description can be used as input to a traditional design rule checker. The filter's processing time was 2 minutes and 19 seconds on a VAX 11-780. The plots include only the metal fabrication layer for clarity.

The filter may be extended to incorporate geometrical and connectivity checks, if the primitive operations are extended in the following way. First, each piece of geometry in every definition is assigned a unique connection number. Transistor components are assigned three numbers, for source, gate and drain. When the primitive operation is invoked, it determines whether one of two cases is present: 1) the two pieces are too close, and 2) if they are too close, the two pieces are connected. When the process is through, two lists of information exist. The first list describes the possible design rule violations, and the second list describes the connectivity information. Then the design rule checker determines whether the pieces of geometry which are possible design rule violations are actually electrically equivalent, i.e. connected.

This approach ignores many design rules since it assumes structural components, i.e. transistors and contact cuts, are design rule correct. This short coming could be alleviated in one of two ways. First, each unique component could be displayed graphically to the designer. People are very good at determining violations given small amounts of information. Second, a corresponding procedure for each type of unique structural component could be written. These procedures would incorporate a check for all possible design rule violations involving the component. The number of routines needed would normally be very small.

The approach taken to design rule checking extracts a hierarchical net list in the process of checking for design rule violations. This net list may also be used to check for connectivity against a net list generated by a hierarchical circuit specification.

CONCLUSIONS

The experimental Design Rule Filter has been used to process a number of designs, with encouraging results. In all cases the remaining geometry was substantially less than the original fully instantiated geometry. The implication of this is that a design analysis tool may use the hierarchy of a VLSI design to greatly reduce the work performed.

The filter implemented placed no restrictions on the hierarchical composition rules. This paper suggests that if a designer defines and uses some limited composition rules, design analysis tools can use the knowledge of the composition rules to minimize the overhead of analyzing a large design. This makes it possible to analyze designs of much greater complexity then might otherwise be possible. More work needs to be carred out exploring the applicability of this type of algorithm to all types of design analysis tools.

ACKNOWLEDGEMENTS

I wish to thank Bob Sproull who originally developed many of the ideas on which this work is based and Ivan Sutherland who introduced me to the ideas. I also would like to express my thanks to the participants in the Silicon Structures Project at Caltech for constant support and ideas, with special thanks to Ed McGrath whose many suggestions are incorporated into this paper.

REFERENCES

Baird, H.S, (1977) Fast algorithms for LSI Artwork Analysis, *Proc of 14th D.A. Conf.*, 303-311.

Baird, H.S., (1977) A survey of Computer Aids for IC mask Artwork Verification, *Proc. IEEE Int. Symp. on Circ. and Sys.*, 441-445.

Hon, R.W., Newell, M., (1980) private communication.

McGrath, E.J., Whitney, T.E., (1980) Design Integrity and Immunity Checking: A New Look at Layout Verification and Design Rule Checking, *Proc of 17th D.A. Conf.*, 263-268.

Mead, C., Conway, L, (1980) "Introduction to VLSI Systems", Addison-Wesley, Reading, Massachusetts.

Rowson, J., (1980) "Understanding Hierarchical Design", Ph.D. Thesis, California Institute of Technology.

Simon, H.J., (1962) The Architecture of Complexity, *Procedure of the American Philosophical Society*, vol. 106, no. 6.

Whitney, T.E., (1981) "A Hierarchical Design Rule Checker", Master's Thesis, California Institute of Technology.

Whitney, T.E.. (1981) A Hierarchical Design-Rule Checking Algorithm, *Lambda*, vol. 2 no.1.

COMPONENTS OF
A SILICON COMPILER SYSTEM

Charle' R. Rupp

Corporate Research Group
Digital Equipment Corporation
Maynard, Massachusetts

INTRODUCTION

A silicon compiler translates a behavioral description of a function into a set of geometric images which can be used to fabricate an integrated circuit that computes that function. A sophisticated silicon compiler has the potential of allowing the design of very large circuits based upon the same methodology that allows modern software compilers to generate very complex programs. This paper describes an experimental silicon compiler system called DEA (DEsign Architecture) which emphasizes the similarities and dissimilarities between hardware (silicon) and software compilation. The goal of this system is to provide a designer with the ability: (1) to describe and verify the behavior of a complex circuit, (2) to explore alternative architectures for a design to achieve the preferred performance, size and power dissipation characteristics, and (3) to generate the geometric description of the circuit for fabrication with a quality comparable to hand drawn designs.

One particular problem which will be referred to as the "side assignment problem" arises when attempting to carry out completely general and optimal hardware compilation. This problem which is fundamentally two-dimensional in nature apparently has no direct counterpart in the normally one-dimensional software compilation process.

SYSTEM MODEL

The approach taken in DEA is to restrict the definition of the software and hardware components of a digital system so that: (1) software is the description of the activities of a machine as a sequence of steps in the time dimension and (2) hardware is the description of the set of concurrent operations in two of the spatial dimensions. Specifically, hardware consists of interconnected switching components. Some of these switching components are connected in such a manner that the signal outputs may assume state values as a function of the order of activation of other signals. These switching components and their interconnect taken together will be referred to as memories. Software defines the values of the memories at each step in a casual manner. The switching components and interconnections which do not form memories will be referred to as combinational circuits. The combinational circuits determine the computations which may occur in a single software step.

The DEA language supports this view of computing by allowing the designer to specify: (1) memory arrays (RAMs, ROMs, and PLAs), (2) combinational relationships which infer switching components and their interconnections and (3) cyclic relationships which define discrete memory components. The DEA language borrows several ideas from the higher level ISP notation (Bell, 1971) and it is hoped that at some point in the future that the two notations may converge allowing an even more powerful facility for the definition and synthesis of digital systems.

OVERVIEW OF THE DEA LANGUAGE

To emphasize the similarities between hardware and software compilation the syntax of the DEA language is a subset of the "C" programming language (Kernighan, 1978). This has been supplemented only slightly by allowing the specification of binary valued variables called "signals" using the conventions of ISP notation. This syntax allows the designer to break up a complex circuit (program) into smaller circuits (procedures) using the concept of a design hierarchy (Whitney, 1981). The circuit procedures or symbols may have signal parameters so that the same symbol definition may be used in several different instances with different signals.

There are two main differences between DEA and its software counterpart: (1) there are no control flow operators such as "if" and "while" defined in DEA (future versions may include this capability to simplify the specification of memory data) and (2) the assignment statement has the meaning of a concurrent relationship between signal variables. The following example illustrates the use of DEA to describe a simple RS flip-flop with a clocked slave section:

```
block FF(Q,S,R,C);
{ signal Q0,Q1,QB;
  { ~Q1=S|Q0;  ~QB=Q|C&Q1; }
  { ~Q0=R|Q1;  ~Q=QB|C&Q0; }
}
```

In this example, Q,S,R, and C represent parameter signals, Q0,Q1 and QB represent local signals that do not cross the symbol's boundaries, the characters "{" and "}" represent the beginning and the end of functional blocks (also called slices), and the characters "~", "&", and "|" represent the Boolean NOT, AND and OR functions. Since the definitions of Q1 and Q0 are circular, these equations define a memory component. There are two basic rules for determining the time execution of a DEA program: (1) signals that are the result of combinational relations change value with zero delay and (2) memory variables change value as a function of their inputs with unit time delay. With this definition, the equation C=~C for example forms an oscillator with a period of two units of time.

Since the equations in DEA are identity relations, the equations may be given in any order without changing the behavior of a circuit. This allows the order of the equations to be used by the designer to imply the relative geometric relationships between the components implied by the behavioral equations. The block structure implied by the begin/end markers forms a "slicing hierarchy" for defining a two dimensional layout of components as follows. The order of equations and references to lower level blocks define a "slice" (one begin end pair) at the current level which implies the order of the components in one dimension. The order of the components in the next lower level block definitions define the order of components in the orthogonal direction. Thus, with two levels of slicing hierarchy, a two-dimensional arrangement of components may be specified. Using several slicing levels provides considerable flexibility for defining a general composition and it is not unusual for a design to have as many as 20 to 30 levels in the slicing hierarchy. The levels at which symbols are instantiated will be referred to as the "calling hierarchy". The levels which are defined within a symbol will be referred to as the "intermediate hierarchy". The inclusion of geometric semantics in DEA allows the designer considerable control over the geometric layout of a circuit as well as allowing relatively efficient compilation procedures.

DESIGNER'S VIEW OF DEA

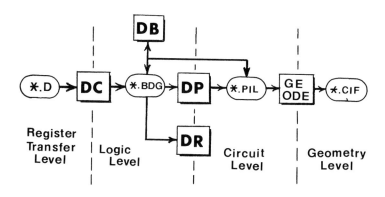

Figure 1. DEA System Components

Figure 1 illustrates a simplified flow of data files and data processing programs in the DEA system. The designer initially prepares a behavioral description of a circuit in the DEA language. In the early stages of design, geometric relationships are normally ignored except for the implications of the calling hierarchy. The DEA definition file (*.D), which may include references to library functions and other useful circuits, is translated into a file (*.BDG) of canonical logic equations by the DEA language compiler program (DC).

The canonical form of logic used in the DEA system is a forest of binary decision trees following the semantics of the Binary Decision Graphs (BDG) of Shelly Akers(1977). This canonical form of logic has been found to be extremely compact in the representation of complex systems and provides the basis for a number of computationally efficient simulation, test vector generation and synthesis procedures. Of particular interest is the property that the pass transistors and dynamic memory nodes commonly found in MOS circuitry can be readily modelled and synthesized.

Once the designer has corrected syntactic errors in the source file the *.BDG file is "executed" by a functional level simulation program (DR). Using this program, the designer may verify the correctness of the description by viewing a variety of waveform and tabular data displays which result from the "running" of the hardware program defined in the *.D file. This is generally accomplished by including test circuits in the source description to assist in the generation of the appropriate time sequences. Pseudo-random number generators and counters are particularly useful for this purpose. These test circuits may be discarded in the geometry generation process. The DR program must perform an ordering operation on the signals to ensure the correct modelling of behavior. The current implementation does this by causing purely combinational variables to be computed in topological order (Aho, 1974) followed by the computation of memory variables in reverse topological order.

Once the designer is satisfied with the behavior definition, floorplan geometry information is included in the DEA source by ordering the behavioral equations and adding intermediate slicing levels. Generally, at least one intermediate level is added for each symbol so that each symbol is a two-dimensional object.

The revised DEA description is then recompiled and processed by the "pillow" generation program (DP). The resulting file (*.PIL) contains the designer's geometric definition along with a translation of the canonical form of logic equations into a set of switch equations. These switch equations indicate the primitive circuit components needed to implement the logic using the technology chosen by the designer. The symbols in the *.PIL file are called "soft" symbols or "pillows" because they have a rectangular shape but arbitrary flexibility in the spatial dimensions.

Once the designer is satisfied with the floorplan of pillow symbol compositions the geometry program (GEODE) is used to harden the soft symbols into hard symbols (fixed geometry). The resulting file (*.CIF) may then be used to control mask generation equipment for use in final circuit fabrication.

GENERATION OF SOFT CELLS

The first level at which any sense of technology and geometry is introduced is in the process of converting the BDG equations to switch equations in the pillow files. To illustrate the process we will assume the availability of primitive component symbols such as a pulldown (PD), pullup (PU) and NOR symbol as illustrated in figure 2. Using these components, the behavioral equations and slicing in the previous flip-flop example imply the layout or "floorplan" and interconnections indicated in figure 3a. Figure 3b illustrates the Akers' diagram for this same function. The intermediate slicing boundaries are shown as dashed lines.

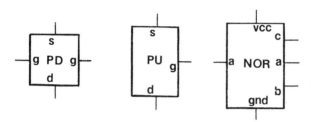

Figure 2. Example Design Components

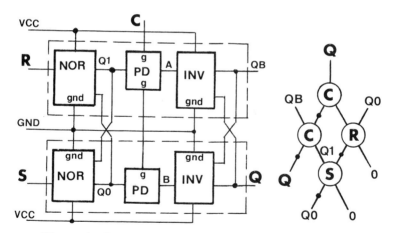

Figure 3. Example Flip-flop Circuit Composition

The DP program performs the following functions as an intermediate step in translating to geometry: (1) vertical and horizontal directions are assigned to the slicing divisions, (2) the connectors (signals which cross a slice boundary) at the calling levels of the hierarchy are assigned to the sides of a rectangle which will be eventually hardened to form the final geometry of each symbol, (3) the BDG equations are translated into technology dependent switch equations, (4) load factors are computed for each signal so that information relating to interconnect wire size and transistor gate size can be determined (extra switch equations for inverting signals and load buffering are also added at this point), (5) the scope of variables is checked to ensure connectivity and determine the global signals at each calling hierarchy level, (6) the area of each symbol is estimated based upon the complexity of the switch equations, and (7) an ideal aspect ratio for the horizontal and vertical dimensions is generated.

The union of the sets of global signals and the parameters to a symbol definition form the set of the symbol's connectors. The ground and power literals ("gnd" and "vcc" in figure 3) which are implied by the use of regular logic equations in the original DEA description are the most common examples of global signals to a symbol. The designer may vary the performance and power dissipation of the circuit to a limited extent at this level by adjusting speed/power ratios and fanout load factors on a global or local basis.

Figure 4 illustrates the set of three-valued partial binary logic operators which are useful in describing the circuit topology for a circuit implemented in the NMOS technology. By changing the values for the "!" operator, switch equations may be formed for circuits implemented in the PMOS and CMOS technologies as well.

A	B	A?B		A	B	A!B		A	B	A:B
0	x	u		0	x	0		0	0	0
1	0	0		1	0	u		0	1	u
1	1	u		1	1	1		1	0	u
u	x	u		u	1	1		1	1	1
								0	u	0
								1	u	1

x = don't care   u = undefined

Figure 4. Switch Operator Function Tables

Generally, the operator "?" represents a pulldown device, "!" represents a pullup device, and the ":" operator is the "WIRED-OR" operator which allows partial functions to be combined to form total logic functions. A simplified pillow file description of the flip-flop example using NMOS technology is as follows:

```
pillow FF(Q,S,R,C); global(gnd,vcc);
right(Q); left(gnd,vcc,R,S); top(C);
vertical; local(Q0,Q1,QB);
  horizontal;
    switch Q1=Q1!vcc:S?gnd:Q0?gnd;
    switch QB=QB!vcc:C?Q1:Q?gnd;
  end;
  horizontal;
    switch Q0=Q0!vcc:R?gnd:Q1?gnd;
    switch Q=Q!vcc:C?Q0:QB?gnd;
  end;
end;
```

Of the operations performed on the design by the DP program the most complex is the assignment of signals to the sides of the conceptual rectangles. The program first orders the symbol definitions by instance-area. Instance-area is defined to be the product of the estimated area of a symbol and the number of instances of that symbol in the overall design. The elements of the intersections between connector sets of neighboring symbols are then added to the appropriate sides of each symbol subject to the restriction that the same signal reference may not appear on adjacent sides of a symbol. Alignment constraints are generated from this information if a butting connection may be possible.

GENERATION OF HARD CELLS

The GEODE program is responsible for translating the pillow symbol definitions into final geometry. Switch equations are converted to hard symbol transistors and additional intermediate slices by recursively applying a set of circuit templates (typical floorplans of NOR, NAND and MEMORY structures). This is followed by a sequence of hierarchy transformations which attempt to optimize the circuit floorplans. Figure 5 illustrates the final hierarchy representation including the two new intermediate slice levels for a portion of the flip-flop example.

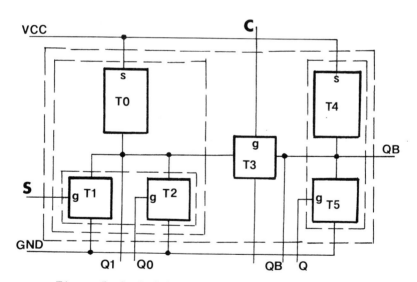

Figure 5. *Switch Equation Intermediate Levels*

The process of bottom-up wiring proceeds as follows. The symbol definitions are again ordered by instance-area (using the revised estimate of area based upon the unwired circuit). For each symbol, the lowest level slices are converted to semi-hard slices by using a general set of wiring procedures to connect the one dimensional sequence of components of that slice. These wiring algorithms have the property that the resulting slice is arbitrarily stretchable in the direction of the slice.

In this process all (except as noted below) components are optimally oriented to minimize interconnect wiring and intermediate slice connectors are formed to allow composition with the next higher level of symbol slices. Then the next higher level slices are wired in a similar manner taking advantage of lower level slice stretching when possible to minimize the overall composition area while at the same time using any alignment constraints which may be present. Once a symbol has been placed in an instance, that symbol is completely hard and may not be stretched further. Once all the symbols have been hardened the final geometry of the desired integrated circuit is written into a geometry file in the Caltech Intermediate Form (CIF) (Mead, 1980) file format. This file is then ready for use in fabrication. Any symbol may be replaced prior to wiring by hardened geometry generated by other means. In particular, the designer would normally replace the symbols with the most instance area by hand designed symbols. Analog circuits for interface pads and memory arrays are incorporated in the same manner.

A FUNDAMENTAL PROBLEM

A particularly interesting problem arises in the use of the completely general composition scheme outlined in the previous section. To fully describe the problem, it will be useful to define three generic types of compositions which occur in the process of hardware compilation: (1) a "leaf cell" composition is any composition in which the majority of the components have fewer than four logical connectors (actually the number varies but four is an average number which happens to correspond to the number of sides of a rectangle), (2) a "butting cell" composition is any composition in which the majority of components are stretchable slices having side alignment constraints, and (3) a "general composition" is any composition which is not a leaf cell composition or a butting composition. These cases occur naturally in the order defined when wiring a circuit from the bottom up.

The fundamental problem of general hardware compilation involves the composition of leaf cells. Although the problem probably manifests itself in several different ways, the manner in which it appeared in the current work has to due with the assignment of logical signals to the sides of the intermediate slicing levels. A circular problem arises as follows. If the correct orientations of the components of a slice are known, then the connector assignments may be made by simply extending the connectors of the components out to the slicing boundary. If the connectors are known on the slicing boundaries then the components of the slice may be optimally oriented to minimize wiring to the boundary. When the number of logical connectors of each component is small and the components are relatively square, simple optimizing metrics such as area and side correlation have a very weak minimum prohibiting the selection of either side connections or component orientations with a high degree of confidence. Figure 6 illustrates this situation for the composition of a NOR gate. An arbitrary selection of orientation and side assignment for these cases invariably creates problems when attempting to complete the wiring at the higher levels. Current approaches for circumventing this problem either restrict the floorplan of the leaf cells (for example, all signal wires to gates of transistors run in the horizontal direction) or incorporate hand drawn leaf cells as for example the leaf cells generated by a system such as described by Mosteller (1981).

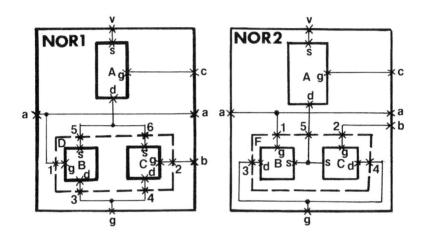

Figure 6. Possible NOR Gate Side Assignments

CONCLUSION

By applying known concepts of software compilers an almost completely general hardware compiler for the generation of integrated circuit geometry can be constructed having the properties of relatively simple user source definition, verifiable functional behavior, reasonable processing time and reasonable quality of result. Complete geometry of good quality can be generated if the ability to replace selected leaf cells by hand drawn definitions is included. The ultimate quality of a silicon compiler appears to be limited by the solution used for the leaf cell side assignment problem. Although restricted leaf cell floorplans may give nearly optimal results, further research on the side assignment problem may give important clues as to how hardware compiler "code generators" may be further improved.

ACKNOWLEDGEMENTS

I gratefully acknowledge the many discussions which contributed substantially to this work with the students, faculty and SSP representatives at the California Institute of Technology. The work reported here was carried out as part of the Silicon Structures Project at Caltech under sponsorship from the Digital Equipment Corporation.

REFERENCES

Aho, A.V., et. al., (1974) "The Design and Analysis of Computer Algorithms", Addison-Wesley, Reading, Massachusetts.

Akers, S.B., (1977) On the Specification and Analysis of Large Digital Functions, *Proceedings of the 7th International Symposium on Fault Tolerant Computing*, pp.88-93.

Bell, C.G., (1971) "Computer Structures: Readings and Examples", McGraw-Hill, New York.

Kernighan, B.W., (1978) "The C Programming Language", Prentice-Hall, Englewood Cliffs, New Jersey.

Mead, C., Conway, L., (1980) "Introduction to VLSI Systems", Addison-Wesley, Reading, Massachusetts.

Mosteller, R., (1981) REST: A Leaf Cell Design System, *VLSI 81 International Conference*, University of Edinburgh.

Whitney, T.E., (1981) A Hierarchical Design Analysis Front End, *VLSI 81 International Conference*, University of Edinburgh.

OVERVIEW OF THE CHiP COMPUTER

Lawrence Snyder

Department of Computer Sciences, Purdue University, West Lafayette, IN 47907 USA

1. INTRODUCTION

There has been a rush to exploit VLSI technology by building special purpose devices tailored to a particular complex algorithm: tree machines for searching, sorting and expression evaluation, systolic array processors for numerical calculations, graph and combinatorial algorithms. The leverage comes from identifying locality (for high integration) and uniformity (for mass production) in the algorithm. But the problem remains:

How does one compose these algorithmically specialized processors into a larger system?

We must put them together to solve more complex problems.
One solution to the composition problem is to attach a variety of these specialized processors to a bus, but the benefits of the devices are lost in wasteful interprocessor data movement. Alternatively, the algorithmically specialized processors could be emulated by microprocessors connected in a perfect shuffle or other general interconnection network, but this wastes enormous area, forgoes locality and introduces routing delays. The devices could be wired together, but this achieves only one composition and the best order for the outputs of one processor is not always the best order for the inputs to the next. The Configurable, Highly Parallel (CHiP) computer permits the composition of algorithms in a way that retains the locality and uniformity of the special purpose devices while providing flexibility.

2. THE CHiP ARCHITECTURE FAMILY

First we present an overview of the main components and function of CHiP computers. In subsequent sections the capabilities and limitations of the components are discussed in detail.

Informal Overview

CHiP architectures are characterized by a switch lattice connecting a set of homogeneous microprocessors (PEs) that is a slave to a controlling sequential computer (the controller). The *switch lattice* is a regular structure formed from programmable switches connected by data paths. The PEs are connected to the switch lattice at regular intervals rather than being directly connected to each other. External storage devices connect to the lattice at the perimeter switches.

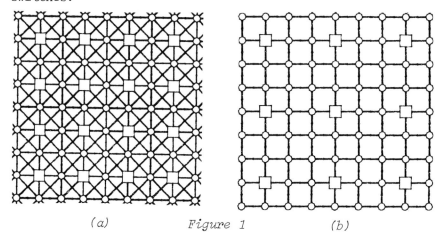

(a)   *Figure 1*   (b)

Figure 1 shows two examples of switch lattices. The switches are shown as circles, the PEs as squares and the data paths as lines. Although the PEs and switches are drawn in roughly the same scale in the figure, the PEs are substantially greater in both area and capability. The examples represent a portion of the lattice that may contain 2^8 to 2^{16} PEs. Current technology permits only a portion of the lattice to be placed on a single chip, but because of the characteristics of the architecture discussed below, "wafer level" fabrication is possible.

The switches are circuit rather than packet switches and each contains sufficient local memory to store several configuration settings. A configuration setting enables the switch to establish a direct, static connection between two or more of its incident data paths. For example, to achieve a mesh interconnection pattern of PEs for the lattice in

Figure 1(a), we assign North-South configuration settings to
the switch rows and East-West settings to the switch columns.
This pattern is illustrated in Figure 2(a). The same lattice
has been configured into a binary tree pattern in Figure 2(b).

As mentioned in the Introduction one motivation for the
CHiP architecture is to provide a flexible means of composing
algorithms to solve large problems. Accordingly, we can
visualize an algorithm as being divided into a sequence of
phases, each with its own interconnection pattern. The PEs
will each be performing different operations in the various
phases as the emulated devices change.

To prepare for a sequence of phases, the controller loads
the switch memories with the proper configuration settings
to achieve the different interconnection patterns. This is
performed by means of a separate interconnection "skeleton"
that is transparent to the lattice. Typically, the loading
of switch settings takes place in parallel with the loading
of program segments for the phase into the PE memories. The
configuration settings for the same phase are loaded into
the same memory location in all of the switches. For example, the settings for a tree could be stored in location 1,
the settings for a mesh in location 2, etc.

On a broadcast command from the controller, all switches
implement the configuration setting in the same location.
With the entire lattice configured, the PEs begin synchronously executing the instructions stored in their local memory
in response to this same broadcast command. PEs need not
know to whom they are connected; they simply execute instructions such as READ EAST and WRITE NORTHWEST. The
configuration remains static until the controller broadcasts
another command causing a different configuration setting to
be implemented. The new interconnection pattern for the next

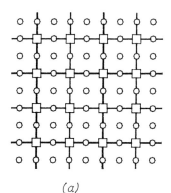

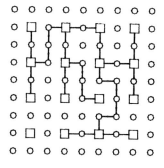

(a)   Figure 2   (b)

phase is established in a simple logical step and PE instruction execution resumes. (Detailed examples of this use of configurability are given in Gannon and Snyder 1981.)

Clearly, members of the family of CHiP architectures can differ in many ways: complexity of the lattice, functional complexity and memory capacity of the PEs, number of PEs, interconnection capability and memory capacity of the switches, width of the data paths, geometry and controller capacity. We next discuss some of these possibilities.

Processing elements

The computational capacity of the PEs largely determine the degree to which a CHiP machine is a general purpose computer and is thus influenced by the intended applications. For example, if the CHiP computer is to be used to simulate the action of other VLSI circuits for design verification purposes, PEs with a few dozen gates suffice to emulate a node of the circuit. Numerical algorithms provide an enormous class of applications requiring a floating-point arithmetic capability and substantial (100-200 instruction) programs. Since complex functions can be implemented in software, our intuition says that memory capacity is more important than functional capability. Early experience with CHiP algorithm design corroborates this view. We have been able to make effective use of a technique Schwartz (1980) calls "summarizing", and therefore we recommend that PEs used in an $n \times n$ lattice have sufficient memory to store at least n data values.

Switches

The operation of the switches is very simple and they might be implemented entirely with "steering" or "pass" transistors were it not necessary to intersperse drivers. Even so, they occupy area that is only a small factor larger than the minimum, m^2, where m is the number of wires of the path pairs. The number c of memory locations for storing configuration settings will be small (<16) and the degree d, which is the number of incident data paths, will be either four or eight.

The crossover capability of a switch refers to the number of distinct data paths groups that can be independently connected by a switch. We refer to "groups" rather than data paths since fan-out is possible, i.e. more than two directions can be connected simultaneously. The crossover number g varies between 1 (no crossover) to $d/2$, and will generally be two since it appears to be difficult to make very effective use of more crossovers.

The total number of bits required at a switch is thus, dcg, one for each direction for each crossover group for each configuration setting, though this number is modest, it can be reduced by permitting settings to be assigned by the controller while the PEs are executing. This "asynchronous loading" capability exploits the fact that configurations often differ in only a few positions.

Lattice

The lattice structure determines the efficiency of PE utilization and the convenience of embedding interconnection patterns. The crucial variable is the *corridor width, w,* the number of data paths separating two adjacent PEs. (Recall that a single data path is formed from m wires.) Figure 1(a) shows a $w=1$ lattice while Figure 1(b) shows a $w=2$ lattice. Both lattices are uniform in the sense that all PEs are separated by the same size corridors, but it is possible to have a lattice with a variety of corridor widths. For example, the lattice of Figure 1(b) could be enhanced by enlarging the width of every fourth row and column of corridor pairs to a width of 4. Such an approach permits the lattice to be interpreted at several "levels" of detail: as an $n/4 \times n/4$ PE lattice with corridors of width 4 composed of logical processors that are 4×4 lattice with $w=2$, or as an $n \times n$ lattice with $w=2$, or by ignoring two of the four added corridors.

To see the impact of a particular choice of corridor width, we must study how the lattice *hosts* an interconnection pattern graph. There are two considerations when hosting a pattern graph: PE degree and edge density of complex interconnection paths. The matter of the PE degree, the number of incident edges, can be dismissed easily. If the pattern graph vertices, which correspond to PEs, have a degree in excess of the four or eight degree PEs typically provided, then we simply couple PEs together to give a larger "logical PE". For example, to adjacent degree four PEs can be logically coupled together to give a degree six PE; one of them could simply act as a buffer. Although this reduces the number of available PEs, the problem arises infrequently, since few processes require such large numbers of operands simultaneously. When few operands must be recieved from many possible sources, the large degree problem can be solved with the fan-in provided by the switches.

The second problem is that the graphs edge density may

require many different data paths to pass through a region of the CHiP lattice. In theory, even a one corridor lattice can host such a pattern, but to do so may require PEs to be unused in the region in order to provide sufficiently many paths. For example, Figure 3 shows an embedding of $K_{4,4}$ into the lattice of Figure 1(b). In order to provide paths for the sixteen edges, the center four PEs must be unused. Increasing the corridor width obviously raises PE utilization.

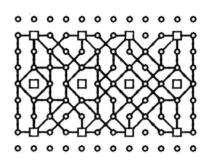

Figure 3

It also lowers the PE density. The fact that the number of PEs in linear in the area of this lattice means that PE utilization is inversely related to the area required to embed the pattern graph in the plane. Graphs requiring a nonlinear area will underutilize PEs just as circuits described by these graphs are composed mostly of wire, (Thompson 1980).

The decision on how wide corridors must be is influenced by the intended interconnection patterns and how economically necessary it is to maximize PE utilization. Fortunately, many algorithmically specialized processors developed for VLSI implementation have linear area interconnection graphs and can be hosted with optimal or near optimal PE utilization when the corridor width is only two. But to hose any planar graph in an $n \times n$ PE lattice, an (average) width proportional to at least $(log\ n)^{\frac{1}{2}}$ will be necessary, (Leighton 1981), to achieve optimal PE utilization. (Valiant, 1981 shows that $log\ n$ width suffices.) For optimal utilization in more complex pattern embeddings such as the shuffle-exchange graph, a much larger width is required (e.g. proportional to at least $n/log\ n$, Thompson 1980). These lattices with nonconstant corridor width have sublinear PE density per unit area.

The impact of these results is as follows.

> Up to constant multiplicative factors, the CHiP lattice with constant corridor width use the silicon area as efficiently as direct VLSI implementations for all pattern graphs; CHiP lattices with constant corridor width as well as such universal interconnection structures as the shuffle-exchange graph cannot use the silicon area any more efficiently than direct VLSI implementation or constant corridor CHiP lattices and they are less efficient for linear area pattern graphs.

Evidently, a constant width corridor is indicated

The estimates of the precious paragraphs are based on asymptotic results involving large constant factors and refer to a purely planar model. They can serve as guidelines (especially for "wafer level" fabrication), but more practical considerations are likely to influence the implemented lattice structure. For example, if only a small portion of the lattice fits on a simple chip, the chips must be wired together which gives an opportunity to implement a complex nonlocal interconnection structure in the "third dimension". "Pin" limitations will also influence the decision. It may be more efficient to use the pins to increase the parallelism of data transmission using wide data paths through a narrow corridor than to use them for a wide corridor of narrow data paths. The benefits would accrue to the linear area patterns.

Efficient Embedding of Interconnection Patterns

Even though an interconnection pattern graph may have a linear area embedding in the plane, a direct translation of that embedding into the CHiP lattice may not be perfectly efficient. For example, the well known "hyper-H" planar embedding of a binary tree (Meand and Conway, 1980), when literally translated into a lattice of corridor width one, leaves nearly half of the PEs unused. The reason is that unlike plain silicon, the CHiP lattice provides predetermined sites for the PEs which must be respected.

It is possible (Snyder 1981) to have a perfectly efficient embedding for the complete binary trees in a lattice for which $w=1$. That is, a $2^k \times 2^k$ PE lattice can host a complete binary tree with $2^{2k}-1$ nodes. Figure 4 illustrates a portion of this intricate, but straightforward, embedding.

Although the embedding of Figure 4 is efficient in terms

of PE utilization, there are other considerations to be weighed. In particular, propagation delay is an important problem and this embedding contains paths of length proportional to n for an $n \times n$ PE lattice. There are planar embeddings for complete binary trees with paths of length proportional to $n/\log n$ (Paterson et al), which is the best possible. These embeddings can be made just as efficient in terms of PE utilization and would probably be preferred.

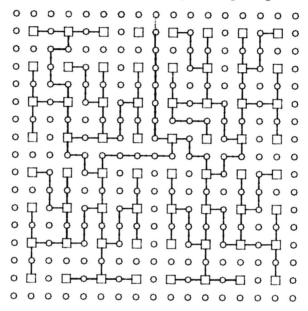

Figure 4

3. CONCLUSIONS

We have introduced the CHiP architectures and discussed some of the ·design decisions influencing this structure. The multiphased processing paradigm was discussed as a means of using configurability to compose algorithms. But there is another way to use configurability to compose algorithms.

Pipelined algorithms are best composed by coupling the PEs of one algorithm with those of the next in such a way that the outputs of the first become imputs to the second. The CHiP processor supports that composition method easily. Each of the algorithms is embedded in a region of the lattice so that the input side(s) of the first region is adjacent to the perimeter and the output side of this and subsequent processor arrays are adjacent to their successor algorithms's PEs. This approach uses "intraphase" configurability (as opposed to the "interphase" configurability of the previously mentioned paradigm) to arrange for the embedding and connectivity of the processor arrays and to "scale" them since their size often depends on a parameter of the input. The controller can make these modifications at loading time. A complete example of composing the Kung and Leiserson systolic arrays (Mead and Conway, 1980) is given in Snyder, 1980.

Obviously, the CHiP architecture is very fault tolerant. Once a faulty PE, switch or data path is discovered, it is a simple matter to configure around the offending elements(s). Perhaps the most effective way to use this facility is to wafer level fabrication. With this approach the wafer is viewed as a enormous chip where the dicing corridors are used for data path corridors. A wafer is accepted if, after testing, a regular $k \times k$ sublattice was found to be functional. Onboard mapping circuitry could map the addresses of the logical PEs and switches onto the functional elements of the wafer.

In summary, the CHiP architecture provides flexibility for algorithm composition. Accordingly, the many design parameters left unspecified in this discussion will be set based on our ongoing research into algorithm design for the CHiP architectures.

ACKNOWLEDGEMENTS

It is a great pleasure to thank my colleagues, Dennis Gannon, Janice Cuny, George Holober, Ching Hsias, Paul McNabb and Kye Hedlund, who have contributed in innumerable ways to these ideas. The work described herein is part of the Blue CHiP project which is supported in part by Office of Naval Research Contracts N00014-80-K-0816 and N00014-81-K-0360. The lattice is Task SRO-100.

References

Gannon, D.B. and Snyder, L. (1981), "Linear Recurrence Systems For VLSI: The Configurable, Highly Parallel Approach," Int'l Conference on Parallel Processing, (to appear)

Leighton, F.T. (1981), "New Lower Bound Techniques For VLSI" Twenty second Ann. Symposium on the Foundations of Computer Science, IEEE (to appear)

Mead, C. and Conway, L. (1980) *Introduction to VLSI Systems* Addison-Wesley

Snyder, L. (1981) "Introduction to the Configurable, Highly Parallel Computer," Purdue University Tech. Report 371

Thompson, C.D. (1980), A Complexity Theory for VLSI, Ph.D. Thesis, Carnegie-Mellon University

Valiant, L.G. (1981) University Considerations for VLSI Circuits IEEE, Transactions on Computers

IMPLEMENTATION OF SLA'S IN NMOS TECHNOLOGY

Kent F. Smith

Department of Computer Science, University of Utah

Salt Lake City, Utah, USA

1. INTRODUCTION

The development of a methodology for the implementation of structured logic arrays in integrated circuits using extensive computer aids is the subject of an intensive research effort by the VLSI group at the University of Utah. The methods being investigated are for the implementation of arrays in three different integrated circuit technologies, I2L , NMOS, and CMOS . The efforts include the circuit design , composite layout, analysis, and where practical , the actual fabrication of the circuits in each of these three technologies. Concurrent investigations are being carried out for the development of general software tools for these implementations . Some of the software being developed includes a data base which can adequately represent the structured logic array, an editor, a simulator, and a placement/routing program to operate on this data base.

The form of structured logic which is being investigated is known as the stored logic array (SLA). The SLA was originally conceived by Suhas Patil [1] and further elaborated on by Patil and Welch [2]. Patil's idea was to build structured logic arrays using a "folded" programmable logic array (PLA) which contains flip-flops distributed throughout the array with arbitrary column and row breaks to give multiple, independent, finite state machines on a single integrated circuit. Circuit design is performed by placing logic symbols on a grid which is the physical representation of the actual chip. This placement performs both the logic description and the interconnection of the integrated circuit while simultaneously giving all of the necessary information for the generation of the composite.

Thus the system designer can visualize the logical description of the system in terms of the physical layout of the IC, a technique which we call "visual perception" of a logical design. This technique offers a new dimension to VLSI design.

2. SCHEMES FOR CIRCUIT DESIGN

Circuit Design Techniques

The circuit design techniques which were investigated for the NMOS SLA all involve synchronous logic with two or more clocks. Previous designs in I2L SLAs [3] used both synchronous and asynchronous logic. The ease of using pass transistors and clocks in NMOS along with the expertise which exists in the design of synchronous circuits dictated the synchronous design methodology for the NMOS SLA.

Two different synchronous schemes were explored for the implementation of the synchronous NMOS SLA. These were (1) dynamic multi-phase clock schemes (four or more) which used very little ratio logic and took advantage of precharged nodes for dynamic circuit operation and (2) static two-phase clock schemes using ratio type logic for all inverters with clocks and pass transistors for clocking signals between different logic levels. The two different implementations have various advantages and disadvantages which can be understood by examination of a specific implementation in each scheme.

Static Implementation

The two-phase static circuit closely approximates the I2L implementation which was previously described. This NMOS circuit uses the same four column wires for the set, reset, Q, and Q' signals from the flip-flop as in the I2L designs and two column wires for the inverter. The row logic functions are accomplished by the use of single NMOS transistors, similar to that used in the I2L SLA.

The operation of the circuit is best understood by examination of the simple SLA shown in Fig. 1. This figure contains the SLA program, the non-overlapping, two phase clock waveforms, and the circuit implementation. This simple SLA does not represent any useful function but rather is used to illustrate the relationship between the SLA symbols and the actual circuit.

Dynamic vs. Static Implementation

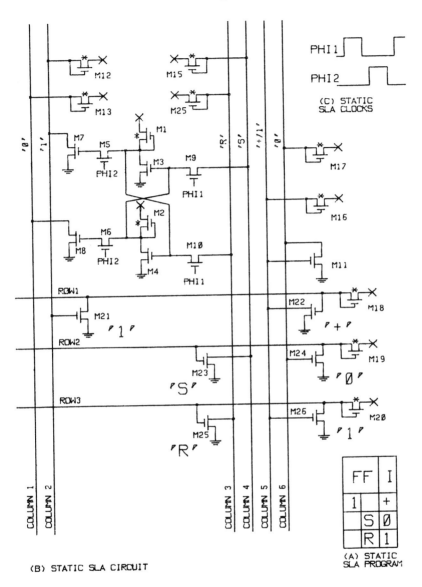

Fig. 1. Static SLA Showing SLA Program, Circuit Schematic, and Two-Phase Clock Waveforms

Two of the major reasons for investigating dynamic circuits for SLA implementation are (1) to lower the power requirements by the use of dynamic precharged circuits and (2) to reduce the physical size of the SLA by time multiplexing the column wires. Dynamic implementation is not simple for the SLA and involves careful consideration of the tradeoffs that can be made. For example, one of the requirements in the design of the SLA is to be able to do at least one or more levels of logic in the AND-OR plane via the inverter columns during a single clock cycle. The static scheme allowed unlimited levels of logic because the AND-OR plane and the inverter columns are composed of ratio type logic inverters which can be continuously cascaded. The clocks have to be sufficiently long to allow the signals to ripple around the entire circuit but this is a relatively simple requirement to meet. Dynamic circuits do not enjoy this luxury since each additional logic level in a dynamic circuit requires additional clocks. Thus some compromise has to be made between the total number of clocks and the number of logic levels which are allowed. Another consideration is the retention of data in the flip-flops for long periods of time. In the static circuit this was simple because the ratio logic will hold data so long as power is applied to the device. In the dynamic circuit, however, an additional clock must be introduced to retain the data. The introduction of additional clocks to solve these two problems can have disastrous consequences because of the physical layout of the SLA itself. The arbitrary placement of flip-flops and row cells in the integrated circuit means that these clocks must always be carried somewhere in the array and thus the overall size may actually increase as a result of using these techniques rather than decrease as one might think.

Circuits that have been designed to date using the two phase static scheme do consume large amounts of power. For example, an SLA containing approximately 150 rows and 100 double wire columns which is capable of operating at a 4 MHz clock rate will consume approximately 250 milliwatts of power. This is completely unacceptable for VLSI circuits which may contain thousands of rows and columns. The physical size of the SLA is also very important if VLSI is to be achieved. Time multiplexing of the column wires under the flip-flops can result in fewer column wires and thus a significant space saving. The static SLA circuit used four column wires for the set, reset, Q, and Q' input/outputs of the flip-flop. These four wires can be replaced in a multiphase clock scheme by two wires which time share the set and Q lines and the reset and Q' lines. This technique of using two wires under the flip-flop has been called a "two wire" scheme as compared to the "four wire" scheme in the static SLA.

Three dynamic circuits were designed using the "two wire" scheme with several combinations of precharged rows and columns with both dynamic and static inverter columns. All three of the circuits used static flip-flops and thus none were completely dynamic. Two of the circuits were limited to four clock phases and one of the circuits was a six phase scheme. The first four phase clock scheme contained precharged rows and flip-flop columns with ratio logic in the inverter columns. Thus it was possible to perform unlimited numbers of logic levels with the inverter columns using this scheme but only one level of logic in the AND-OR plane. The second four phase scheme used the same precharged rows and flip-flop columns as used in the first scheme as well as a precharged inverter column. This scheme only allowed for a single logic level with the inverter column. The six-phase scheme was similar to the second four-phase scheme except that the circuit actions were not time shared as was the case with the four phases.

Dynamic Implementation

An explanation of each of these three different circuits exceeds the space limitations of this paper. However, an examination of the circuit for the first four phase scheme described above will provide some insight into the techniques which were used in the design. This circuit along with an SLA program and the four phase clock waveforms is shown in Fig. 2. The operation of the SLA program is identical to that described in Fig. 1 for the two-phase scheme. There is only a single logic column (two column wires) under the flip-flop cell as opposed to the double logic column (four wires) under the flip-flop in the two phase scheme. The inverter column is identical. The flip-flop is implemented using two static ratio type cross-coupled logic gates. The flip-flop columns and rows are both precharged but the inverter column uses ratio type logic.

A decision was made to continue the investigation of the SLA with the static scheme and to leave the dynamic circuits for further investigation. The dynamic circuits are still the only solution to the power problem in the NMOS technology and eventually this type of approach must be taken. However, the most pressing issue is not the power issue but rather the methodology of the SLA and the potential use of this concept for the design of structured logic. The reasons for the choice of the static over the dynamic can be summarized as follows: (1) Multiple levels of logic in the AND-OR plane and the inverter columns are easily accomplished with the static circuit. (2) The dynamic circuits were actually larger than the static

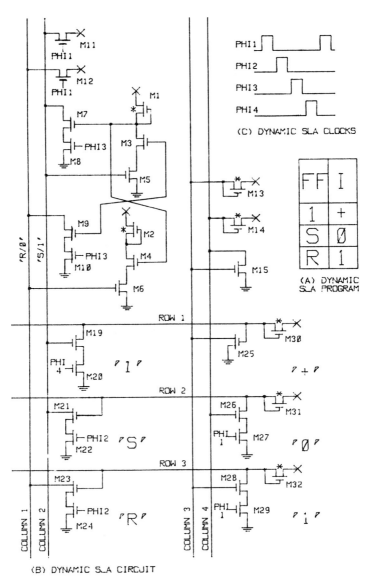

Fig. 2. *Dynamic SLA Showing SLA Program, Circuit Schematic, and Four-Phase Clock Waveforms*

circuits. There was a significant space savings in the "two wire" system but the double transistors in the row elements actually increased the overall size of the circuit. (3) The additional clocks complicated the power distribution in the SLA which increased the size of the circuit and (4) The circuits were still static because of the flip-flops and in some cases the inverters. Thus the power was still a significant factor.

3. STATIC SLA CELLS

A complete set of cells for the static SLA were designed and composite drawings of each cell were made. The cells included the following: (1) Four combinations of read and write enable R-S type flip-flops, (2) Four combinations of read and write enable D type flip-flops, (3) Two inverter cells for odd and even columns, (4) Eighteen row and column load cells, (5) Seven odd and even column "1", "0", "R, "S", and "+" cells, and (6) Fifteen miscellaneous interconnect cells for column connect, row connect, end caps, and blank rows.

The process used for the design of the cells was a seven mask (including glass overlay) Silicon Gate NMOS process with five micron minimum feature size. The process included the use of a buried contact between the polysilicon layer and the N diffusion. Both enhancement and depletion type transistors were used in the design.

The basic grid structure of the SLA was dictated by the dimensions of the logic row cells which include the "0", "1", "S", "R", and "+" cells. These functions all occupy a single column width and a single row height. Composite layouts of these cells resulted in a row cell which was 35 microns high and 75 microns wide. The R-S and D flip-flop cells were five rows high and two columns wide. The inverter cell was the size of a single row cell. Each row and column segment is terminated by a row or column load cell. There are several row and column load cells with different width and length ratios. The selection of an appropriate load transistor is dependent on the required performance of the SLA.

4. EXAMPLE OF SLA IMPLEMENTATION

An example of an SLA program for the implementation of a pre-settable four bit up-down counter using the two phase static scheme will now be discussed. The SLA program for this counter is shown in Fig. 3 . The counter is composed of four flip-flops , FF1 - FF4 in columns 11 - 18. A fifth flip-flop, FF5 in columns 19 - 20, is used for edge triggering of the counter. There are 10 inverters, I1 - I10 in columns 1 -10 which are used to generate the true and the not true values of the input data and the internally generated data. The counter can be preset by the action represented in rows 8 - 15. Each of the four inputs, 0, 1, 2, and 3, columns 5 - 8, are loaded into FF4, FF3, FF2, and FF1 respectively by means of the "1" - "S" and "0" - "R" in rows 8 - 15. The load action can take place only when the load input is a "1".

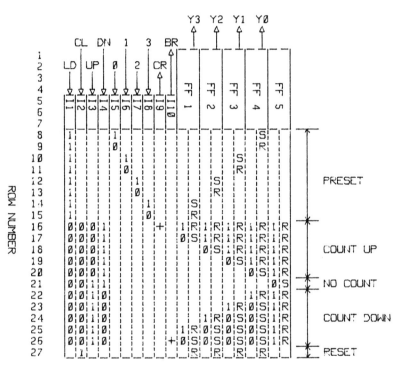

Fig. 3. *SLA Program for Four-Bit Up-Down Counter*

The controls for counting up are contained in rows 16 - 20. The controls for counting down are contained in rows 22 - 26. Counting takes place when the LD (load) and the CL (clear) inputs, columns 1 and 2, are low, "0". Up counting occurs when the DN input, column 4, is held at a "1" and the UP input, column 3, is toggled from a "1" to a "0". Likewise, down counting is accomplished when the UP input is held at a "1" and the DN input is toggled from a "1" to a "0". No counting occurs when both the UP and the DN (down) inputs are a "1". FF5 is used to force a single count whenever the toggling of either the UP or the DN input occurs. Because the counter is operated with a two phase clock, if FF5 were not there, it would count synchronously with the clock whenever the counter was enabled. FF5 is set via row 21 whenever both the UP and the DN inputs are high. Then at each occurance of a low on either the UP or the DN inputs, FF5 is reset and no counting can occur until the UP and DN inputs both go back to the "1" state which causes FF5 to set. The CR (carry) and the BR (borrow) outputs, columns 9 and 10, from the counter are generated by the two "+" cells which have been inserted in row 16, column 9 and row 26, column 10 respectively.

The space occupied by this counter scheme can be significantly reduced by rearranging the rows and columns and by making use of row and column segmentation. Fig. 4 represents the same counter as that shown in Fig. 3 where the cells have been compacted into a much smaller space. This actual cell placement uses 18 columns and 21 rows as compared to the 20 columns and 27 rows in Fig. 3. Row and column pull-up cells and blank cells for carrying signals to various sections of the SLA have also been added. This drawing shows all of the interconnecting row and column wires as hash marks on the bounding cell boxes.

The complete composite of this circuit was prepared and

BNK	BNK	BNK	2PU.B	2PU.B	2PU.B	2PU.B	2PU.B	2PU.B	2PU.B	2PU.B	2PU.B	2PU.B	BNK	BNK	BNK	BNK	BNK
BNK	BNK	BNK	1PU.B	1PU.B	1PU.B	1PU.B	1PU.B	1PU.B	1PU.B	1PU.B	1PU.B	1PU.B	BNK	BNK	BNK	BNK	BNK
BNK	BNK	BNK											BNK	BNK	BNK	BNK	BNK
BNK	BNK	BNK											BNK	BNK	BNK	BNK	BNK
2PU.B	2PU.B	2PU.B	FF		FF		FF		FF		FF		2PU.B	2PU.B	2PU.B	2PU.B	2PU.B
1PU.B	1PU.B	1PU.B											1PU.B	1PU.B	1PU.B	1PU.B	1PU.B
INV	INV	INV											INV	INV	INV	INV	INV
1	BNK	1	BNK	BNK	BNK	BNK	BNK	BNK	RPU.C	RD+	BNK	BNK	BNK	BNK	BNK	BNK	BNK
Ø	BNK	1	BNK	BNK	BNK	BNK	BNK	BNK	RPU.C	S	Ø	RD+	Ø	1	1	Ø	RPU.B
BNK	1	1	BNK	BNK	BNK	BNK	RPU.C	RD+	Ø	RD+	1	S	Ø	1	Ø	Ø	RPU.B
BNK	Ø	1	BNK	BNK	BNK	BNK	RPU.C	S	1	S	1	S	Ø	Ø	1	Ø	RPU.B
1	BNK	1	BNK	BNK	RPU.C	RD+	Ø	RD+	1	S	1	S	Ø	1	Ø	Ø	RPU.B
Ø	BNK	1	BNK	BNK	RPU.C	S	1	S	Ø	RD+	1	S	Ø	Ø	1	Ø	RPU.B
BNK	1	1	RPU.C	RD+	Ø	RD+	1	S	1	S	1	S	Ø	1	Ø	Ø	RPU.B
BNK	Ø	1	RPU.C	S	1	S	Ø	RD+	Ø	RD+	1	S	Ø	Ø	1	Ø	RPU.B
BNK	BNK	BNK	Ø	RD+	1	S	1	S	1	S	1	S	Ø	1	Ø	Ø	RPU.B
BNK	BNK	RPU.B	1	S	1	S	1	S	1	S	1	S	Ø	1	Ø	Ø	+
BNK	BNK	RPU.B	Ø	RD+	Ø	RD+	Ø	RD+	Ø	RD+	1	S	Ø	Ø	1	Ø	+
2PU.B	2PU.B	RPU.B	1	S	Ø	RD+	Ø	RD+	Ø	RD+	1	S	Ø	Ø	1	Ø	2PU.B
1PU.B	1PU.B	RPU.B	BNK	S	BNK	S	BNK	S	BNK	S	BNK	S	1	BNK	BNK	BNK	1PU.B
INV	INV	1	RPU.B	BNK	BNK	BNK	BNK	BNK	BNK	BNK	BNK	BNK	Ø	BNK	BNK	RD+	INV

Fig. 4. Cell Placement for Four-Bit SLA Counter

the circuit is being processed in the Hedco Microelectronics Laboratory at the University of Utah. The chip is 60 X 90 mills and contains 16 pins.

5. CONCLUSIONS

One design methodology for an SLA using NMOS technology has been demonstrated. This design involved the investigation of both static and dynamic NMOS circuits. A two-phase static circuit was used as an example showing the actual implementation of a simple SLA. Future work in SLA methodology will involve design in other MOS technologies as well as further development of computer aids for automating the design process.

6. ACKNOWLEDGMENTS

This work was supported in part by a contract from General Instrument Corp. R. and D. Center, Chandler, Arizona, and in part by a grant from the National Science Foundation, MCS-78-04853. The author gratefully acknowledges the help of his colleagues at the University, particularly Suhas Patil for initial contributions to the SLA concept, Tony M. Carter for his assistance in writing the SLA program for the counter example, and Wing Hong Chow for his assistance in preparing composite drawings. Also the help of William Knapp and William Dunn of General Instrument Corp. was most valuable in defining the dynamic SLA circuits.

7. REFERENCES

[1] S. S. Patil, "An Asynchronous Logic Array, "*Tech. Memo TM-62*, Project MAC, MIT, Cambridge, Mass., May 1975.

[2] S. S. Patil and T. A. Welch, "A Programmable Logic Approach for VLSI," *IEEE Transactions on Computers*, September 1979, Volume C-28, Number 9.

[3] K. F. Smith, "Design of Stored Logic Arrays in I2L, " *IEEE International Conference on Circuits and Systems*, April, 1981, Chicago, Ill.

A PHYSICAL DESIGN MACHINE

Se June Hong, Ravi Nair and Eugene Shapiro

IBM T. J. Watson Research Center, P. O. Box 218, Yorktown Heights, New York 10598, U. S. A.

1. INTRODUCTION

Rapid advances in fabrication technology have made it possible to place tens of thousands of electronic circuits on chips measuring no more than a quarter of an inch per side. There are indications that the level of integration will keep getting higher at least in the near future. There remains, however, a gap between this sophisticated technology and the design tools available to take advantage of this technology. Physical design refers to the technique by which the design of a digital system represented by interconnected logic gates is converted to lithographic patterns to be used in fabricating a chip or a set of chips for the system. Tools for physical design generally take the form of software packages aiding the various steps of the conversion. The three conventional separate, though by no means unrelated, steps are ***partitioning, placement*** and ***wiring*** (Breuer, 1972). Partitioning refers to the process by which portions of a large system are allocated to different chips when a single chip cannot contain the entire design. The placement process places the components of a chip, commonly gates, within the chip. Often, the positions where these components may be placed are a set of well-defined slots, separated from each other by areas through which wires or metal segments may be run in order to interconnect the components. When such a structure takes the form of a rectangular array of slots, it is called a ***gate-array*** or a ***master-slice*** chip. Figure 1 illustrates such a chip. The algorithms used to allocate the space between the components for the interconnection are contained in the wiring software package.

The gap between physical design tools and the advancing technology arises due the fact that the performance of the algorithms for partitioning, placement and wiring is not a linear function of the number of components on the chip but rather gets increasingly worse as the chip complexity increases. Algorithms which perform better in speed generally are less successful in producing good layouts, especially when the percentage of utilised space on the chip increases.

One way of improving the speed of algorithms without sacrificing wirability is to use faster machines on which the algorithms are executed. It is easy to wait for a new generation of high speed general purpose computers so as to

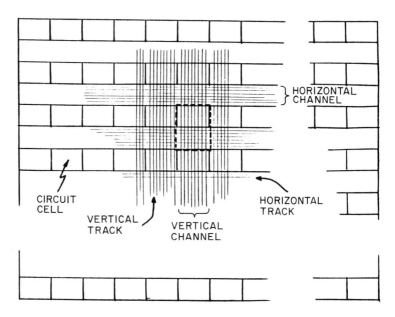

Figure 1. A typical master-slice chip

execute the existing programs faster. It is more effective, though, to design a machine whose architecture takes advantage of some of the special characteristics of physical design algorithms. This will be illustrated by considering the wiring process.

2. THE MACHINE CONFIGURATION

Consider an array of computers arranged in a 2-dimensional matrix of size $n \times n$. Each computer consists of a processor, a memory unit, one port to each of its adjacent computers (four, except at the boundaries) and a link to an additional computer which we shall call the **controller.** An illustration of such a system appears in Figure 2, where the X- and Y-select lines make it possible for the controller to access any desired **node** or some subset of nodes in the array. Further let us assume that the memory of each node computer carries the complete information about a corresponding circuit cell in an $n \times n$ sized master-slice chip. The portion enclosed within the dotted line in Figure 1 typifies a cell on the chip. While such an aggregation of computers would have been impractical just a few years ago, the advent of microprocessors makes such a scheme inexpensive and practical today. There are two ways how such an architecture can vastly improve the performance of wiring algorithms. First, the presence of a computer at each node makes it possible to exploit parallelism in the wiring algorithm. Second, frequently used functions can be hardwired as special instructions in the processing element of each node computer.

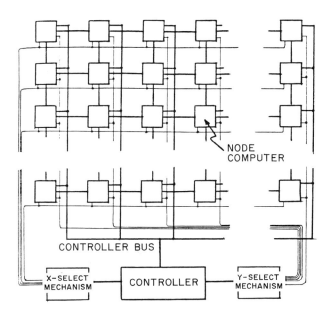

Figure 2. *System configuration for an array of computers*

An experimental version of such a system is operational at the IBM Thomas J. Watson Research Center. It is basically an 8×8 array of computers, each of which consists of a commercial 8-bit microprocessor, 2 kilobytes of memory and auxilliary circuit enabling neighbour and controller communication. The controller is also an 8-bit microprocessor with 48 kilobytes of memory, floppy and hard disk units, printer, terminal and a link to the host computer system of the facility. The following paragraphs detail the steps in performing the wiring. We will describe later how an 8×8 array can be programmed to wire a chip having more than 64 components.

3. GLOBAL WIRING

Global wiring (Chen *et al.*, 1977; Shiraishi and Hirose, 1980) is the name given to the phase of wiring in which wires are allocated to channels in the chip, without specifically assigning tracks within the channel. The latter assignment is accomplished subsequently by an ***exact embedding*** algorithm. Before starting global wiring, each node computes the total number of tracks available in each of four directions. It is also provided with a list of nets which have a terminal in that node. (A *net* is a set of two or more terminals which have to be connected together.)

(a) For each net to be wired, whether it has a terminal in that node or not, a node computes the likelihood of it being a member of some connection in that net in each direction, and adds an appropriate cost to the *congestion estimate* in that direction. The congestion estimate is a decreasing function of the distance of the node from a terminal node of the net.

(b) After step (a) has been completed for all nets, steps (c) through (g) outlined below are performed for the first net, and then iterated, net by net.

(c) For a given net, a node subtracts its own contribution to the congestion estimate for that net from the total remaining congestion estimate. It does this for each of the four ports in turn.

(d) Based on the number of unused tracks and the updated congestion estimate at each port, the node computes *port costs* for each of the four ports. This cost is a measure of the penalty for occupying a track in that direction. Thus, the penalty for going through a port which has only one remaining track will be heavier than one with many remaining tracks. This penalty alone cannot serve as the port cost because it penalises a port with few remaining tracks even if there are not many remaining nets contending for use of that port. The needs of unwired nets are represented in the congestion estimate at the port. Hence a function of the remaining capacity together with the remaining congestion estimate serves better as a measure of port cost. Figure 3 shows the function that is currently being used by the algorithm. The knee of the function occurs at a point where the anticipated channel requirement (demand) equals the remaining channel capacity (supply).

(e) Now comes the *forward propagation* phase. One of the terminal nodes of the net is called the *source* and all the others are called *sinks*. The source, in each of four directions, sends the port cost for the port in that direction to its neighbour in that direction. At the same time, all other nodes send an arbitrarily large value to its neighbours. At each subsequent time step, every node picks up the values coming in at its ports and determines the minimum of these values. A pointer is set to the direction from where this minimum came. For each of its ports the node adds the port cost as computed in (d) to the minimum value and sends the modified value to the neighbour in that direction. Once the node has sent out a meaningful value, i.e. a value that is not the arbitrarily large value described above, it resumes sending this arbitrarily large value at subsequent time steps.

(f) When some sink receives a meaningful value, it stops the forward propagation process by informing the controller. Since it is possible for more than one sink to be reached at the same time, the supervisor, either by polling the cells or by some other scheme, determines the sink cell that has the least minimum value. The *backtrace* now begins. The sink prompts the neighbour in the direction of its stored pointer. At each subsequent time step, a node that has just been prompted by its neighbour prompts another in the direction stored by it during the

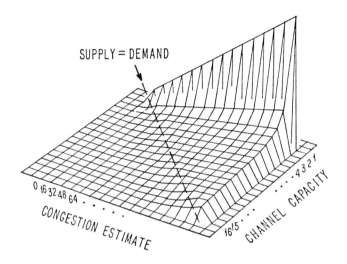

Figure 3. Port cost as a function of channel capacity and congestion estimate

forward propagation phase. This process continues until the source is reached. Any prompted node decrements the channel capacity in the appropriate ports and records the entry and exit points against that net.

(g) All nodes which have recorded entry/exit points for the current net become sources for the next forward propagation. All net terminals remaining to be connected act as sinks. The process for a net terminates when no sinks remain for a net.

The process just described determines the shortest path between the source and any sink, choosing the path with the least cost if there is more than one. It achieves this in time proportional to the rectilinear distance between the source and the nearest sink. When the placement is good, this distance grows very slowly as the chip size increases. In fact, a reasonable estimate for this distance even for a 10,000 master-slice circuit chip is 5. However the number of nets to be wired is approximately a linear function of the chip size, leading us to conclude that the time for the entire process as described is approximately a linear function of the chip size.

There are situations when the shortest path does not lead to the path with least cost. We will call any deviation from the shortest path between a source and a sink as a ***detour.*** A better performance of the algorithm, from the wirability point of view, is obtained when one performs a ***deferential detour.*** In this concept, the nets which are laid down early are not necessarily laid in shortest paths. Rather, they are allowed to detour whenever they reach an area which should preferably be avoided so that future nets do not get blocked in, i.e., congestion due to wiring is anticipated and avoided. This can be achieved by the following modification to the above procedure.

4. IMPROVING THE ALGORITHM

Assume that the path with the least cost is desired, with a detour no larger than d. During the forward propagation phase, each node, instead of quitting (sending the arbitrarily large value to its neighbours) after sending one meaningful value, now quits after sending $d + 1$ meaningful values. (Note that a blockage at a port could cause the port cost itself to be very high. As a result, if no sink is reached, the net is kept aside for wiring later. However, once a node starts sending meaningful values, it is not possible for it to send a higher value at any of the subsequent d time steps.) Each node also maintains a table indicating the direction from which it received the best value against the time step at which that value arrived. If there is a choice of directions, the direction at the previous step is preferred, whenever possible. (This simply helps in getting the shortest path whenever there are paths of different lengths having the same minimum total cost.) The backtrace process is initiated by the sink with the minimum cost prompting the neighbour in the direction stored against its last active time step, say t. While prompting its neighbour, it also sends the value $t-1$ to the neighbour. The neighbour in turn refers to its table and prompts *its* neighbour in the direction stored against the time step $t-1$, along with the value $t-2$. This process continues until, as before, the sink is reached.

Simple modifications to the above procedure can improve the time and space complexity of the above algorithm slightly. For brevity, these are not being indicated here. It is also possible to penalise turns or changes in wiring direction (**via** usage) by maintaining one table for each of the four directions and computing the best cost (which now includes a **via cost**) and best direction for each outgoing port during forward propagation. The algorithm currently uses the maximum number of vias and the minimum number of vias allowed in a cell of a master-slice chip to determine the via cost as suggested in the paper by Lee *et al.* (1981). Again, details are omitted here.

The above algorithm was tested using a small example using the experimental machine described above. The time taken was less than 2 seconds of real time. The program connected all the nets in the minimum Steiner connection when the detour limit, d was set to 0. On increasing the detour limit, the nets took longer paths, but the channel capacity remaining at the end of the run showed a uniform spread over the entire chip. This demonstrates the fact that when nets are allowed to detour, they do so in such a way as to make the wiring of future nets easier. Figure 4 shows the remaining capacity at the cell boundaries for the cases when $d = 0$ and $d = 4$. Note the absence of boundaries with zero remaining channel capacity in the latter case. This proves to be quite helpful for the exact embedding phase which follows global wiring. It was also observed that at the end as the channel demand reduced due to fewer remaining nets, the nets were wired with minimum Steiner length.

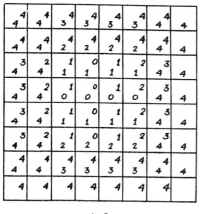

Figure 4. Remaining channel capacity at the cell boundaries for a small example, using detour limits d=0 and d=4 respectively (Initial channel capacity was 4 at all boundaries)

5. FOLDING

In order to make the machine more useful in wiring larger chips, small hardware modifications were made. The edges of the computer array were connected in a wrap-around fashion, so that the nodes at the north edge had the nodes at the south edge as their north neighbours, the nodes at the east edge had the nodes at the west edge as their east neighbour, and so on. Further the memory at each node was increased to 7 kilobytes.

There are two basic ways in which one could attempt to wire, say, a 24×24 circuit chip using an 8×8 computer array. In the first method, one simply partitions the bigger chip into larger partitions so that a 3×3 sized circuit array is represented in each node of the computer array. In the second, which we shall call *folding,* the circuit partition is retained in the 24×24 form. However, various 8×8 *frames* of the chip are worked on at different points in time by the physical 8×8 sized computer array. It may be clear that the wrap-around connections allow every circuit cell in any contiguous 8×8 portion of the chip to be represented by distinct node computers in the array. The folding process performs better than the first method because it takes advantage of the relatively high locality in the distribution of node computation activity without sacrificing the resolution of the wiring process. A node computer must retain the circuit data for every eighth circuit cell in the horizontal direction and in the vertical direction. This suggests the reason for the increased memory requirement at each node. An example for a smaller array is shown in Figure 5. (It should be clarified here that the process described here is more accurately a **cut-and-stack** process, rather than a strict accordion-type folding process. The latter process, while feasible, and while free of the wrap-around hardware requirement, carries with it the inconvenience of alternating directions for consecutive frames.)

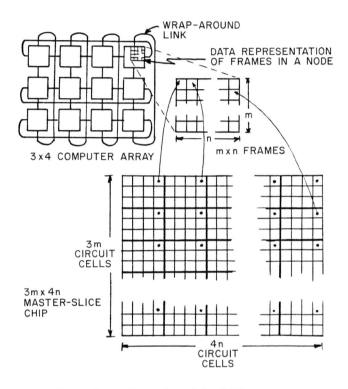

Figure 5. *An illustration of the folding process*

Extensive analysis of the folding process indicates that it is a highly cost-effective process from the point of view of system hardware and turna-round time. The time required to do forward propagation for a 24×24 chip on an 8×8 array machine is not 9 times as much as the time on a 24×24 array machine, but only about 4 times as much. This factor deteriorates to about 8 as the detour limit is increased to 10. Some examples were executed on the experimental machine using data on chips having 19×23 circuits. The real time for execution of the global wiring process for 293 nets increased from 46 seconds with $d = 0$, to 160 seconds with $d = 10$. The analysis also indicates that since nets tend to be rather local even on large chips, a machine with 32×32 node computers would be no worse than twice as slow as an arbitrarily large machine for a large VLSI chip for all reasonable detour limits. It must be remembered though, that the memory requirements at a node computer increases linearly as the number of frames being represented at the node. Figure 6 shows the time penalty incurred by folding a very large chip (more than 10,000 circuits) for various physical array configurations. The effect of detour limit on the penalty is also depicted. (Note that there is no time penalty for the backtrace process which is inherently sequential.)

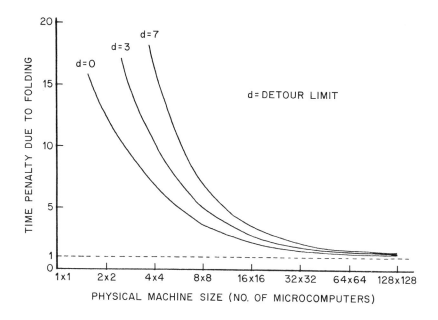

Figure 6. *Time penalty for folding vs. number of nodes in computer array for a VLSI chip*

6. EXACT EMBEDDING

We conclude the description of the wiring process with a brief outline of the exact embedding process which follows the global wiring process. In this phase, each node must allocate exact tracks to the nets which have been assigned to it by the global wiring phase. In order that the assignments be consistent (a track assigned at one circuit cell to a net must be the same as the track assigned to the net at the adjacent circuit cell), it is necessary for the nodes to communicate with each other. Our algorithm begins by making a complete assignment for the circuit cell in the centre of the chip. The tracks assigned to the nets which cross the boundary of the cell are sent to the appropriate neighbouring elements, which accomodate this data in making assignments for its circuit cell. This process continues in a diamond shaped wavefront form until the tracks at the circuit cells at the corners of the chip have been allocated. While details of this process will not be described here, it should be mentioned that extensive lookahead ensures that the track assigned by one processor for a net will be one that is reasonably preferred by the adjacent unassigned cell. A wavefront propagation in the opposite direction, i.e. from the corner cell to the centre cell, has been incorporated at the end of the algorithm to accomodate any conflicts in the assignment.

7. CONCLUDING REMARKS

We believe that a machine of the type described will help in speeding up and improving the quality of other VLSI physical design processes. A larger machine with, say, 32×32 computer elements, where each element is a special purpose chip or set of chips, could improve the speed relative to our experimental system by a factor of 50 or more. This will provide a real-time feedback from the various physical design processes, thus enabling better system design. The extension of the hardware and algorithms for the chip to the physical design of cards and boards is also natural. It is needless to say that the structure of the described machine is suited for other calculations in areas like pattern recognition, image processing, matrix manipulation and finite element techniques.

REFERENCES

Breuer, M. A. (1972). "Design Automation of Digital Systems: Theory and Techniques." Prentice-Hall, New York.

Chen, K. A., Feuer, M., Khokhani, K. H., Nan, N. and Schmidt, S. (1977). The Chip Layout Problem: An Automatic Wiring Procedure, *Proceedings of the 14th Design Automation Conference,* New Orleans, Louisiana, pp 298-302.

Lee, D. T., Hong, S. J. and Wong, C. K. (1981). Number of vias: A Control Parameter for Global Wiring of High Density Chips, *IBM Journal of Research and Development, 25* (to appear).

Shiraishi, H. and Hirose, F. (1980). Efficient Placement and Routing Techniques for Master-Slice LSI, *Proceedings of the 17th Design Automation Conference,* San Diego, California, pp. 458-464.

SESSION 6

OPTIMALITY in VLSI

Bernard CHAZELLE and Louis MONIER

Department of Computer Science
Carnegie-Mellon University
Pittsburgh, Pennsylvania 15213

1. Introduction

The more realistic model of computation introduced in recent papers by Chazelle and Monier (1981a, b) has led to drastic revisions of VLSI complexity in general. Measured in terms of chip area and computation time the complexity of several problems has been shown to be much higher than previously thought. We propose here to investigate the actual performance of well-known circuits in this new model, and to suggest designs which meet criteria of optimality. We will show in particular that many complicated schemes falsely believed to be efficient can be advantageously replaced by simpler and higher performance designs.

Throughout this paper we will base all analyses of circuits on the model of computation described by Chazelle and Monier (1981b). The major new assumption of this model is to require propagation times at least linear in the distance. To make the model suited for upper bounds, we will add that linear propagation time is actually realistic with current technologies, e.g., electrical (with use of repeaters) or magnetic-bubble. Note that, based on these assumptions, the concept of optimality is meaningful only for large circuits, since the actual complexity of very small chips is overshadowed by parasitic effects. However, asymptotically optimal circuits which are conceptually simple will provide useful insights and guidelines for small designs.

One major consequence of the model is that the time performance of a circuit is strongly dependent on its geometry rather than its topology. In particular, all the tree-based schemes previously proposed cease to have their claimed logarithmic complexity. Examples of such circuits can be found in Preparata and Vuillemin (1979), and in Thompson (1980a). Also, to be realistic, we must assume that clock signals follow the same law of propagation as any other signal. For example, broadcasting a control signal in constant time becomes impossible, which significantly alters the control of a device as simple as a shift register.

In this paper, we carry out these ideas and present optimal designs for the following problems: Addition, cyclic shift, integer product, matrix arithmetic, linear transforms and FFT, sorting, and searching. Although we also establish a number of lower bounds, most of those used to prove optimality can be found in Chazelle and Monier (1981b). A common point for all these problems is to have an $\Omega(N^{1/2})$ lower bound on the time T, where N is the size of the problem. This shows the importance of pipelining computations in order to increase the throughput, and leads to the introduction of the period P, defined as the minimum time elapsed between two consecutive inputs. Thus we can analyze the behavior of circuits in the light of three measures: the time of computation T, the period P, and the chip area A. Composite measures defined as products of A, T, or P will also be considered to show possible resource trade-offs.

2. Addition

For the problem of adding two N-bit integers, the following lower bounds have been shown by Chazelle and Monier (1981b): $T = \Omega(N^{1/2})$, $AT = \Omega(N)$, $ATP = \Omega(N^2)$.

The simplest adder, consisting of a full-adder cell, has unit area and runs in linear time. It is thus optimal for the measures A, AT and ATP.

In order to achieve optimal time performance, the carry-look-ahead scheme had been previously proposed. However, it is no longer optimal in our model, for its straightforward implementation yields linear time, while the H-tree layout creates enormous difficulties for ordering the inputs. Instead, we propose a new adder which can be regarded as a one-level CLA. Wlog, assume that $N = m^2$. We decompose each input number $a = a_{N-1}...a_0$ and b into m blocks of m consecutive bits. Each block is read in sequentially into a cell which computes the sum of the two *slices* as well as the last carry, and which checks whether all the bits in the sum are equal to 1. Thus for each block, the sum $(s_{i,m-1},...,s_{i,0})$, the last carry c_i, and one bit $p_i = s_{i,m-1} \wedge ... \wedge s_{i,0}$ are computed. It is important to notice that $c_i = 1$ implies $p_i = 0$.

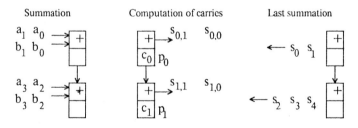

Fig. 1. *An optimal adder.*

After m steps, we can propagate each carry at the block level by computing the actual carry sequentially: if $p_i = 0$ the running carry is c_i, else it is c_{i-1}. Finally we can update the partial sums in $O(m)$ time with the value of the running carry (Fig.1).

With $A=O(N)$ and $T=O(N^{1/2})$, this adder is optimal for T and ATP. Moreover, it requires very little logic since it is made of $O(N)$ shift-register cells and $O(N^{1/2})$ full-adder-like cells. We observe that this scheme seems especially well-suited for magnetic-bubble technology.

One shortcoming of this adder, though, is its failure to allow pipelining. It is easy, however, to design a very simple adder with period $P=1$. Fig.2 illustrates this new scheme. Note that both time and area are linear, while the unit period makes this adder optimal for P and ATP.

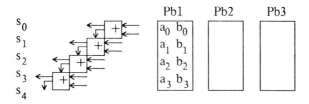

Fig. 2. *A fully-pipelined adder.*

3. Cyclic shift

The complexity of the cyclic shift has been studied by Chazelle and Monier (1981b) and Vuillemin (1980); the following lower bounds have been established: $A=\Omega(N)$, $T=\Omega(N^{1/2})$, $AP^2=\Omega(N^2)$.

The function takes for input a pair (a,p), where a is a binary sequence $a_1,...,a_N$ and p is the value of the shift, and it returns the sequence $b_1,...,b_N$ with $b_i = a_{i\text{-}p[\text{mod } N]}$. We will describe a circuit optimal for both A and T, which is based on a clock-free implementation of a shift register. For simplicity we will assume that $N=m^2$, and that we are given α and β, with $p=\alpha m+\beta$ ($\beta<m$).

Since we assume that α and β are given in binary representation, the first task of the shifter will be to decode these integers. So, we start by describing a decoder, that is, a circuit receiving an integer $i\leq m$ as input and producing a sequence of i 1's. Let $i=i_k...i_0$ with $k=\lfloor \log_2 m \rfloor$.

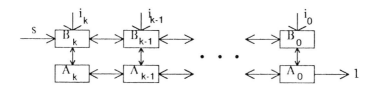

Fig. 3. *A decoder.*

The circuit is a ladder of cells as shown in Fig.3. First the cells $B_k,...,B_0$ read in the bits $i_k,...,i_0$ respectively. Then when the start signal s is activated, if $i_k=0$, B_k

activates B_{k-1} and the process iterates; otherwise it activates A_k, whose task is to generate 2^{k-1} 1's on the output port of A_0. To do so, A_0 simply outputs a 1 when activated by A_1 or B_0, and it acknowledges A_1 in the former case. Recursively A_i activates A_{i-1}, waits to be acknowledged by A_{i-1}, repeats this operation, and when finished, acknowledges B_i. Finally B_i activates B_{i-1} and the process iterates. Each A_i and B_i is a very simple one- or two-state automaton whose details we can omit. Clearly the time to decode any integer $\leq m$ is $O(2^k) = O(m)$.

Next we need to describe a scheme for shifting a chain of bits to a given position without resorting to a broadcast synchronous clock. Consider a chain of m cells $A_1,...,A_m$ with the first p cells holding values $k_1,...,k_p$. We wish to transfer k_p to $A_m,..., k_1$ to A_{m-p+1}. To do so, we simply activate the first p cells in time $O(p)$, then while k_p proceeds to move towards A_m, each cell A_i which is currently holding a key conditionally transfers its content to the cell A_{i+1} if it is vacant. The complete transfer is thus completed in $O(m)$ time.

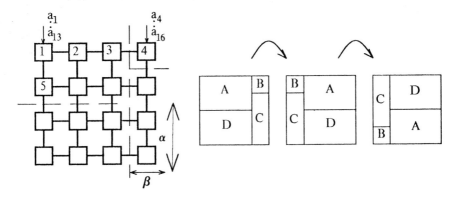

Fig. 4. *An optimal cyclic shifter.*

We are now ready to implement the cyclic shift. The circuit is an m×m array of cells as depicted in Fig.4. The input $a_1,...,a_N$ is read in by rows, and in a first stage the decoder described above permits us to delimit the areas A,B,C,D in $O(m)$ time. Then the actual shift proceeds in two phases as indicated in the picture. This can be done by having two channels between adjacent cells, enabling 2-way communication between them. Then it is straightforward to implement the two phases with the description of the shift register given above.

4. Integer Product

The following bounds are known for the product of two N-bit integers: $A = \Omega(N)$, $T = \Omega(N^{1/2})$, $AP^2 = \Omega(N^2)$. Not only tree-based schemes (Wallace trees) no longer yield logarithmic time, they cannot give better than linear time, since they all generate N^2 temporary bits. Instead, we propose a simple revised version of the shift-and-add scheme (Fig.5). It makes use of two clock-free shift registers meant to

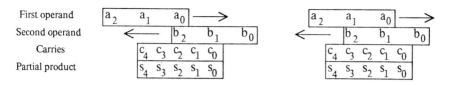

Fig. 5. *The shift-and-add multiplier revisited.*

multiply the two numbers in carry-save representation; the carries are then released sequentially. This circuit is interesting mainly for its simplicity and the fact that it uses minimal area, $A = O(N)$. It is possible to connect N copies of this circuit in order to obtain minimum period, $P = 1$; after each shift-and-add step, operands and partial results are shifted up one position, which after N steps yields the product in carry-save representation. The overall performance is: $A = O(N^2)$, $T = O(N)$ and $P = O(1)$, which makes the circuit optimal for P and AP^2.

Of all the circuits previously held as being *optimal*, e.g., Wallace tree or multiplication through DFT, none shows good time performance in our model. Recently, Preparata (1981) has given a careful description of a fast multiplier, optimal in our model: $A = O(N)$ and $T = O(N^{1/2})$. It uses a square-mesh structure to compute the convolution of the two numbers (considered as vectors) via the DFT, then releases the carries to obtain the usual product. The processors used in this mesh are very similar to those used in the sorting or DFT circuits mentioned in this paper: they mainly are small arithmetical units containing stored programs.

5. Matrix arithmetic

We next turn our attention to three matrix problems: Integer or boolean multiplication, transitive closure, and inversion. Since the first problem is reducible to the last two, as remarked by Savage (1979), the results of Chazelle and Monier (1981b) show that all three problems have the lower bounds, $A = \Omega(N)$, $T = \Omega(N^{1/2})$, with N measuring the total number of elements in the matrices.

Kung's systolic multiplier, described in Mead and Conway (1980), is thus optimal for A and T, and is the best known yet, although it cannot be pipelined. The same remark applies for the mesh-connected circuit proposed by Kung, Guibas, and Thompson (1979) to compute a transitive closure (note that for these results, we assume unit cost for element operations). Similarly Kung, in Mead and Conway (1980), describes an optimal circuit for inverting matrices which admit LU-decompositions (and Gaussian elimination without pivoting). It is easy to extend this scheme to compute the determinant of such matrices in time $O(N^{1/2})$. This matches the lower bound for arbitrary matrices, as is shown in the following.

Lemma 1: The time required to compute the determinant of an arbitrary m×m matrix is $\Omega(m)$.

Proof: Using a result from Chazelle and Monier (1981b), we simply have to show that computing a determinant involves a fan-in on $\Omega(m^2)$ elements.

Consider the matrix A_k defined by the recurrence, $A_1 = (a_{1,1})$

$$A_k = \begin{bmatrix} A_{k-1} & & & r_1 \\ & & & \vdots \\ & & & r_{k-1} \\ a_{k,1} & \cdots & a_{k,k-1} & 0 \end{bmatrix}$$

where the r_i's are the sums of A_{k-1}'s rows, i.e., $r_i = a_{i,1} + \ldots + a_{i,k-1}$. Noting that we can rewrite A_k as

$$A_k = \begin{bmatrix} A_{k-1} & & & 0 \\ & & & \vdots \\ & & & 0 \\ 0 & 0 & \cdots & 0 & 1 \end{bmatrix} \times \begin{bmatrix} I_{k-1} & & & 1 \\ & & & \vdots \\ & & & 1 \\ a_{k,1} & \cdots & a_{k,k-1} & 0 \end{bmatrix}$$

we derive the relation

$$\text{Det}(A_k) = -\text{Det}(A_{k-1}) \times (a_{k,1} + \ldots + a_{k,k-1})$$

which proves that the assignment $a_{i,j} = 1$ for all i,j; $1 \leq j < i \leq k$, gives a hard input, that is, an input for which a change in any one of these assignments alters the value of the determinant. This shows that computing $\text{Det}(A_k)$ involves a fan-in on $k(k-1)/2 + 1$ elements, which completes the proof. □

6. Linear Transforms and Discrete Fourier Transform

Vuillemin (1980) has shown that any circuit which can compute any linear transform on N k-bit elements (with $k \geq \log_2 N$) computes a transitive function of degree Nk. Using this result, lower bounds have been found for this problem by Monier and Chazelle (1981b): $A = \Omega(Nk)$ and $T = \Omega(N^{1/2}k^{1/2})$. One possible implementation requires that, in a first stage, a description of the transform be passed to the circuit as a parameter, thus enabling the circuit to compute the transform on any input vector. As yet, the best method known consists of either storing the matrix on the chip in its traditional array representation, or treating it as part of the input. In both cases, we can then use a matrix-vector multiplier, such as the linear-time, linear-area, systolic multiplier proposed by Kung in Mead and Conway (1980). This circuit is not optimal in time, but it seems hard to improve its performance as long as the N^2 elements of the matrix are to be memorized.

Note that we are often interested in computing only specific linear transforms. The previous lower bounds no longer hold, since they yield no information on the behavior of a particular transform. We choose to turn our attention to one of the most important, the discrete Fourier transform. For this problem, a lower bound is already known: $AT^2 = \Omega(N^2k^2)$. We extend this result in the following.

Lemma 2: Any circuit computing a DFT on N k-bit numbers requires area $A = \Omega(N)$ and time $T = \Omega(N^{1/2}k^{1/2})$.

Proof: The DFT is computed in the ring of integers modulo M. Let ω be a Nth root of unity in this ring. Necessarily $M > N$; moreover we suppose that $k = \lfloor \log_2 N \rfloor + 1$. The DFT is defined by $Y = MX$, where $X = (x_0, \ldots, x_{N-1})$, $Y = (y_0, \ldots, y_{N-1})$, and the matrix M is (ω^{ij}), for $0 \le i, j < N$.

Noticing that the first element y_0 is the sum of all the x_i, we can prove that one of its bits is a fan-in of $O(Nk)$ input bits, simply by exhibiting a *hard-input* (see Chazelle and Monier (1981b) for definition of fan-in). Let $x_0 = 2^j - 1$ with $j = \lfloor k/2 \rfloor$, and $x_i = 0$ for $i = 1, \ldots, N-1$. The jth bit of $y_0 = 2^j - 1$ is equal to 1; however, a change in the value of any bit of order $\le j$ of any x_i ($i > 0$) will force it to 0. Hence, this particular bit is a fan-in of $Nk/2$ input bits, which yields the desired lower bound on the time.

To prove the result on the area, we show that the circuit must memorize at least N bits. Since the order in which the bits are input is fixed, consider the bit b input last, with b being the sth-order bit of x_j. Any y_i can be written as $y_i = a_i + \omega^{ij} 2^s b$, where a_i is independent of b. It is clear that changing the value of b affects the value of all the y_i's, since $\omega^{ij} 2^s$ cannot be zero modulo M.

It follows that, at the instant which just precedes the reading of b, at least one bit of every y_i cannot have been output. Since the DFT is invertible, these bits must be able to take on arbitrary values, which implies that the circuit must memorize at least N bits. □

A good survey of DFT circuits can be found in Thompson (1980b). Note that most of the schemes reviewed are optimal in a model where transmission costs are neglected. However, this ceases to be true in our model, where no logarithmic times are possible. Good examples are the straightforward implementation of the FFT network, or the CCC-scheme proposed by Preparata and Vuillemin (1979), where wires of length N contribute to the poor performance: roughly $T = O(N)$, $A = O(N^2)$.

Instead, we can see how very simple circuits are indeed optimal in our model. One simple method consists of performing a matrix-vector product, where the matrix (ω^{ij}) is input to the hexagonal systolic multiplier. We have $A = O(N^2 k)$ and $T = O(Nk^{1/2})$ provided that optimal adders and multipliers are used. This circuit can be pipelined, thus reducing the period to $O(k^{1/2})$, which is optimal for the measure AP^2.

Another solution consists of using a linear matrix-vector multiplier, generating the matrix elements *on the fly*, as described in Kung and Leiserson (1979). The circuit is more efficient, $A = O(Nk)$ and $T = O(Nk^{1/2})$, but it cannot be pipelined.

A near-optimal design is the square mesh used to simulate a FFT network, as described by Stevens (1971) and Thompson (1980a). It involves N processors, each having the complexity of a microprocessor. The size of the programs involved, however, is comparable to the length of the words processed, which makes the total area $O(Nk)$. Note that the computation time $T = O(N^{1/2}k^{1/2})$ is optimal and matches the I/O time.

7. Sorting

From Chazelle and Monier (1981b) we know that any circuit sorting N k-bit numbers, with $k \geq 2\log_2 N$, requires $\Omega(N)$ area and $\Omega(Nk)^{1/2}$ time. Only the grid implementation of sorting networks proposed by Thompson and Kung (1977) appears to give optimal time performance. However, this scheme requires $\Omega(Nk)$ area, and thus does not match the known lower bound. This lower bound does not exclude the existence of a bit-serial sorting device, but no optimal-area scheme has yet been proposed.

8. A general data structure scheme

As of yet, only the priority queue described in Leiserson (1979) has the complexity initially claimed. All the tree-like structures fail to be logarithmic for reasons already stated. Simple geometric considerations show that it is impossible to access any element of a set of N keys in less than $N^{1/2}$ steps. However we can achieve this performance with the scheme illustrated in Fig.6. The circuit we propose can be used as a dictionary where queries are answered in $O(N^{1/2})$ time, with unit period.

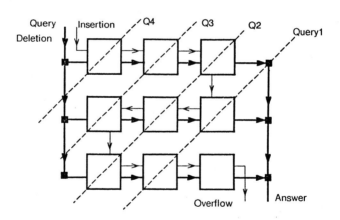

Fig. 6. *An optimal data structure.*

Insertions of keys proceed from the single input port by scanning the mesh sequentially in snake-order, and detecting the first vacant cell. Deletions and queries are handled by a downward diagonal sweep. Thus by using broadcasting, deletions are completed in $O(N^{1/2})$ time. Queries are treated in a similar fashion, running on rows in a left-to-right motion, and downwards on the rightmost column where the partial results from each row are merged. Proper synchronization is crucial; in particular the treatment of insertions and queries/deletions must occur in an alternate mode. Note that at all times we can distinguish between two kinds of keys: the keys stored at their final place, and the running keys in search of a vacancy. However, deletions and queries treat both of them in a similar way. A rigorous proof of correctness is left to the reader.

9. Conclusions

As our last application of the geometric nature of the model considered here, we can show how the idea proposed by Browning in Mead and Conway (1980) to solve NP-complete problems must now be discarded. The method is to simulate a non-deterministic Turing machine by using an exponential number of processors connected up in a tree structure. The paths between root and processors contain a polynomial number of nodes; however, a simple geometric argument shows that for any embedding of the tree, there exists paths of exponential length. This implies exponential communication times, which defeats the purpose of the scheme.

In this paper we have wished to show how the advantages of the high concurrency offered by VLSI technology should be appreciated from a realistic perspective. So far it appears that simplicity and elegance in design should win out. Complicated schemes don't seem to pay off, and even if this observation has been justified in an asymptotic -thus highly academic- framework, we still believe that it bears great significance on a more practical level. After focusing on lower bounds at length, theoretical work in VLSI should now turn decidedly to more practical aims including design and conception.

Acknowledgments

Many thanks to Mike Foster who read a preliminary version of this paper, and to Franco Preparata for communicating to us his recent results on multiplication. This research was supported in part by the Defense Advanced Project Agency under contract F33615-78-C-1551, monitored by the Air Force Office of Scientific Research.

References

Chazelle, B.M. and Monier, L.M. (1981a). *Towards More Realistic Models of Computation for VLSI*, proc. of Caltech Conference on VLSI, January 1981.

Chazelle, B.M. and Monier, L.M. (1981b). *A Model of Computation for VLSI with Related Complexity Results*, proc. 13th Annual ACM Symposium on Theory of Computing, ACM, May 1981.

Guibas, L.J., Kung, H.T. and Thompson, C.D. (1979). *Direct VLSI Implementation of Combinatorial Algorithms*, proc. Caltech Conference on VLSI, January 1979.

Kung, H.T. (1979). *Let's Design Algorithms for VLSI Systems*, proc. Caltech Conference on VLSI, January 1979.

Kung, H.T. and Leiserson, C.E. (1979). *Systolic Arrays for VLSI*, Sparse Matrix Proceedings 1978, pages 256-282. Society for Industrial and Applied Mathematics, 1979.

Mead, C. and Conway, L. (1980). *Introduction to VLSI Systems*, Addison-Wesley, 1980.

Preparata, F.P. (1981). *A Mesh-Connected Area-Time Optimal VLSI Integer Multiplier*, Technical Report, Coordinated Science Laboratory, University of Illinois at Urbana-Champaign, March 1981.

Preparata, F.P. and Vuillemin, J. (1979). *The Cube-Connected-Cycles: A Versatile Network for Parallel Computation*, proc. 20th Annual Symposium on Foundations of Computer Science, Oct.1979.

Savage, J.E. (1979). *Area-time Tradeoffs for Matrix Multiplication and Related Problems in VLSI Models*, Proc. 17th Annual Allerton Conference on Communications, Control, and Computing, 1979.

Stevens, J.E. (1971). *A Fast Fourier Transform Subroutine for Illiac IV*, Technical Report, Center for Advanced Computation, Illinois, 1971.

Thompson, C.D. (1980a). *A Complexity Theory for VLSI*, PhD dissertation, Department of Computer Science, Carnegie-Mellon University, 1980.

Thompson, C.D. (1980b). *Fourier Transform in VLSI*, Memorandum No. UCB/ERL M80/51, University of California, Berkeley, October 1980.

Thompson, C.D. and Kung, H.T. (1977). *Sorting on a Mesh-Connected Parallel Computer*, C.A.C.M., April 1977, v.20, No.4.

Vuillemin, J. (1980). *A Combinatorial Limit to the Computing Power of VLSI Circuits*, Proc. 21st Annual Symposium on Foundations of Computer Science, Oct. 1980.

CHIP BANDWIDTH BOUNDS
BY LOGIC-MEMORY TRADEOFFS

Robert H. Kuhn
Department of Electrical Engineering & Computer Science
Northwestern University
Evanston, Illinois 60201

1. INTRODUCTION

This paper explores the limitations of chip architecture imposed by increasing the scale of integration. IC designers need to quantify several design parameters; to date most *VLSI complexity* results are limited to *time* and *area*. This paper explores the design constraints imposed by *pins*. Although there is no doubt that packaging technology could increase the pins per package, controlling the use of pins will always enhance reliability. In addition, scaling arguments show that pads and pad drivers do not scale as favorably as time (in gate delays) or area. This paper explores the relationship between *pins* and: *area-time bounds, memory intensive design* and *processor to memory area* for specific problems.

In the first section, we present bandwidth, or data rate, as an abstract representation of pin complexity. Pins are used in message switching modes (as in a multiplexer) as well as to transmit computational and control data. In addition, pins used to transmit computational and control data may be time multiplexed. For these reasons pin counting can be deceptive. Bandwidth is put forth as a more accurate, abstract measure of the input/output requirements of a VLSI design. The bandwidth of a design may be measured instantaneously or averaged over time. This paper considers primarily the average input bandwidth.

In the second section, theoretical bounds on bandwidth are explored. Previous research has considered the bandwidth bounds imposed by active logic circuits. These results support Rent's rule and have the form: $B_w = O(A^{\frac{1}{2}})$ where B is the bandwidth of the design (interpreted as pins for Rent's rule) and A is the chip area used for interconnect (interpreted as gates for Rent's rule). This paper takes a different approach, we explore the implications of *memory intensive design* on bandwidth. This is explored by proving lower bounds for bandwidth in terms of memory and processing elements for specific problems. The bounds developed hold for a wide range of memory and processing elements, $N \geq M \geq P$ where N is the problem size, M is the amount of memory, and P is the number of processing elements. The tools used to establish these bandwidth bounds are pebbling games and interconnection networks. For example, the bandwidth lower bound derived for the FFT has the form: $B = \Omega(NP / (N + M \log M))$. Such results are unanticipated if one considers only Rent's rule or its VLSI complexity counterpart. This result establishes a bandwidth-memory (or bandwidth-area) tradeoff which we argue captures the essence of the memory intensive design style.

The last section of the paper explores architectural issues involving bandwidth, time and area. The paper shows that these bandwidth bounds are tight over much of the design space. We discuss trading off processing area for memory area to achieve minimum time. Finally, we mention the value of other bandwidth reducing techniques.

1.1 Bandwidth As a Measure of Pin Complexity

To avoid implementation dependent issues such as multiplexing of pins we propose to measure the *bandwidth* of a design, not the pins. Bandwidth, sometimes called data rate, is a measure of the bits input or output by a circuit per unit time. We define the *instantaneous input bandwidth* at time step t of a computation to be

$$B(t) = I(t)/\tau$$

where $I(t)$ is the number of words input in the t th time period of duration τ. In general input bandwidth is a sufficient measure of total bandwidth because inputs are usually more numerous than outputs. In addition, once an output is computed it either leaves the system forever or it re-enters the system at some later time. In the latter case, it will contribute to the measured input bandwidth. A more theoretically tractable quantity is the *average input bandwidth* hereafter referred to simply as *bandwidth* defined as

$$B = I/T$$

where I is the total number of data input and T is the effective computation time of an algorithm. We use the term *effective computation time* to encompass both sequential, parallel and pipelined designs.

1.2 Design Styles and Fundamental Bandwidth Relations

We start by observing that there are two fundamentally different relationships between bandwidth and area in VLSI designs. They are:
1) A Rent's rule type relation (explained below) in which bandwidth is an *increasing* function of area. We call this relation a *rho-bandwidth relation*.
2) A locality rule type relation which bandwidth is a *decreasing* or constant function of area. We call this relation a *lambda-bandwidth relation*.

Lambda relations have the form of a bandwidth-area tradeoff, in contrast to a rho-bandwidth relation in which bandwidth increases with area. In practical designs, rho relations are encountered when the designer speeds up the system by adding more *parallelism* either at logic or architecture level of the design. A lambda relation is generally encountered when the designer incorporates more *memory* into the design whether it is at the logic level by using memory intensive design techniques or at the architectural level with cache, instruction buffers, or microprogramming. Figure 1 illustrates the effect of these two relations in terms of the gate-to-pin ratio. These relations form an envelope within which all designs lie. The lambda relation is the

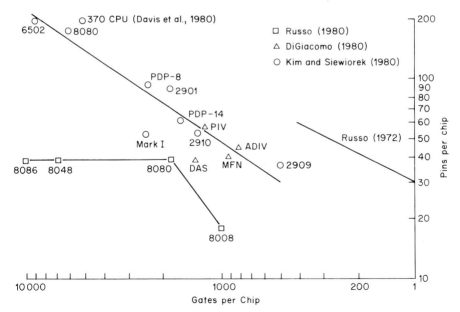

Fig. 1. The Gate-to-Pin Ratio

lower bound on pins while the rho relation is the upper bound. Microprocessors have followed the lambda relation curve; while high performance and data path designs have followed the rho relation curve. The success of memory intensive design techniques in producing lambda-bandwidth relations lead us to adopt it as a major pin conservation technique. In the next section, we will demonstrate a theoretical basis for this phenomenon. In the meantime, a brief discussion of known results concerning the rho relation will be useful.

1.3 Rho Bandwidth Relations

The rho-bandwidth relation is the better understood of the two relations. Landman and Russo (1972) performed experiments verifying a conjecture by Rent that when one averages over a design $P \approx f\, G^r$ for some $0 < r \leq 1$ where P is the number of pins, f is the average fanin of a gate, and G is the number of gates on a MSI to LSI chip. They found that r was bounded by $.57 \leq r \leq .75$ for the designs they collected data on. Another interesting result was that high speed designs tended to have larger values of r. This hypothesis has recently been tested by Kim and Siewiorek (1980) on VLSI data path chips. Finally, although Rent's rule has not been tested at the multiple processor scale of integration, it appears likely to hold by analogy.

On the theoretical side, some bounds on rho relations have been discovered. Vuillemin (1980) has derived lower bounds on area in terms of bandwidth for a class of problems. His results coalesce various earlier area-time tradeoff studies from the perspective of bandwidth. The problems in this class include what he calls transitive functions including: cyclic shifts, integer product, linear transforms, and matrix product. He is able to show that any circuit computing a transitive function satisfies:

$$A \geq a_g N + a_p B + a_w B^2$$

where A is the chip area, N is the number of inputs, B is the average bandwidth, for constants: a_g, a_p, a_w. This formulation subsumes previously derived bounds on area-time of the form $AT^2 \geq aN^2$. The terms in Vuillemin's bound represent gate area, pin area and wire area respectively. It is interesting to point out that Rent's rule is theoretically supported by this result if one considers the wire area term, $a_w B^2$, but it is not supported by the gate area term,

$a_g N$. This result reinforces observations that interconnect area is a primary cost in VLSI designs. Below however, it is shown that one can hope that even better bounds can be attained for bandwidth-area curves by using design techniques obeying the lambda relation.

2. A PROCEDURE FOR GENERATING BOUNDS ON CHIP BANDWIDTH

As engineers, we constantly seek rules by which we can bound our search through a design space. In this section, a procedure for bounding the bandwidth of a design via architectural parameters (memory, processors, and time) is established. For a theoetical basis, we argue that if memory intensive design is the means for generating the more preferable lambda-bandwidth relations and if bandwidth bounds are sought, then it is natural to explore the computation complexity field of pebble games. (Pippenger (1981) surveys this field.) The procedure established is also based on modelling computations as abstract interconnection networks. This allows abstraction from the particulars of any one solution to general bandwidth bounds.

2.1 A Parallel Pebble Game

The *pebble game* simulates the effect of limiting the memory space used by a computation on the time to perform the computation. Several results have shown that for the computation studied there is a tradeoff between time and memory; insufficient memory to store all intermediate results leads to recomputation increasing computation time. To apply the pebble game to bandwidth bounds for VLSI architectures it would be more accurate to augment the pebble game to include parallelism. This would permit one to model the parallelism inherent in VLSI chips.

In the pebble game, the computation is modelled by a *directed acyclic graph* (DAG). The goal of the game is to move a set of identical pebbles, M, from the input vertices of the DAG to the output vertices. In addition to M pebbles, we assume that $P \leq M$ pebbles can be placed on each move (P represents the number of processing elements.) In this game each move is as follows.
1) Zero or more pebbles may be removed from their positions on vertices.
2) Up to P pebbles may be placed on a set of vertices which had all of their input vertices pebbled at the start of the move.

To add a large degree of realism to this model, it is further assumed that $P \leq M \leq N$, where N is the problem size parameter.

2.2 Lower Bounds on Inputs

To establish problem not algorithm specific bounds, we consider only the connectivity of the problem. This leads to the characterization of problems in terms of *"abstract" interconnection networks*. An abstract interconnection network is a DAG. And, it can be shown that any computation of specific problems when modelled as a DAG must contain an abstract interconnection network of a certain connectivity. Perhaps the best known abstract interconnection network is an N-superconcentrator.

An *N-superconcentrator* is a DAG with N-inputs and N-outputs such that for any k ≤ N, for any set X of k inputs, and for any set Y of k outputs, there exists a set of k vertex-disjoint paths from X to Y.

To establish bandwidth bounds for a specific problem, we first establish a lower bound on the number of inputs needed to parallel pebble the abstract interconnection network contained in the problem. In the case of an N-superconcentrator we first establish the following lemma.

Lemma 1: For an N-superconcentrator with M of its vertices pebbled, to parallel pebble any set of M + Q outputs, 1 ≤ Q ≤ P, requires N-M inputs to be pebbled.

Proof: Observe that to be able to pebble a set of outputs, Y, there must be no pebble-free paths from any input of the DAG to any output in Y. Let X be the set of inputs with a pebble-free path to at least one Y. The assertion is that $|X| \geq N-M$. Assume to the contrary that $|X| \leq N-M-1$. Thus there exists at least M + 1 inputs that do not have a pebble-free path to any element in Y. But then M pebbles disconnect sets of size at least M + 1 which contradicts the definition of an N-superconcentrator since vertex sets disconnected by M vertices are connected by M vertex disjoint paths (Menger's Theorem). □

Now to parallel pebble an entire N-superconcentrator, we can show:

Corollary: Parallel pebbling an N-superconcentrator requires

$$I \geq N + (N-M)\lfloor(N-P)/(M + P)\rfloor.$$

Proof: Consider the set of outputs pebbled first. It could contain from 1 to P outputs by the rules of the parallel pebbling game but it requires at least N inputs to be pebbled. Now consider sets of outputs in the order in which they are pebbled. Choose the sets to be of a size as close to M + 1 as possible. (Each could contain up to M + P in the worst case.) There are at least $\lfloor(N-P)/(M+P)\rfloor$ of them. By Lemma 1 each of these requires N-M inputs. The result follows. □

These proofs follow the style of Tompa (1978) where he demonstrates time lower bounds for conventional pebbling. We have shown that his results can be extended to parallel pebbling when $P \leq M$ with only minor degradation.

The problem is now reduced to showing that any computation of the problem of interest contains an abstract interconnection network of a certain connectivity complexity. Several results in this area have been established. For example, consider the problem of computing the convolution, $C(x)$, of two polynomials $A(x)$ and $B(x)$.

Lemma 2: Any bilinear program for computing the convolution, $C(x)$, of two N-1 degree polynomials, $A(x)$ and $B(x)$,

$$c(i) = \sum_{j=0}^{N-1} a(j)b((i-j) \bmod N)$$

where

$$A(x) = \sum_{j=0}^{N-1} a(j)x^j, \quad B(x) = \sum b(j)x^j, \text{ and}$$

$$C(x) = \sum_{j=0}^{N-1} c(j)x^j,$$

contains an n-superconcentrator.

Proof: See Valiant (1976). (The proof proceeds by a linear independence argument.) □

2.3 Upper Bounds on Time

Having a lower bound on inputs via Lemma 1, its Corollary, and Lemma 2, an upper bound on the time needed to compute a convolution will establish a lower bound on bandwidth. Because parallelism is limited in the systems under consideration, namely $P \leq M \leq N$, a variety of efficient parallel algorithms are available. However, a potential algorithm may be complicated by the memory limit.

For example, to compute a convolution, we can use the familiar Fast Fourier Transform theorem

$$C = F^{-1}(F(A) \cdot F(B)).$$

The convolution is found by taking the inverse Fourier transform of the product of the Fourier transformed polynomials. An algorithm for computing a FFT with limited memory and multiple but limited processors can be derived by extending the algorithm of Savage and Swamy (1978) to $P > 1$ processors. We can show:

Lemma 3: The FFT of two N-1 degree polynomials, $N = 2^n$, can be computed in

$$T \leq O((N^2/M_f - N \log M_b)/P)$$

with memory $M = M_f + PM_b$ and P processors where

$$M_f = 2^m, \quad M_b = n - m, \quad \text{for } 0 \leq m \leq n, \text{ and}$$

$$P = 2^p, \quad \text{for } 0 \leq p \leq m.$$

Combining Lemmas 1 through 3 we can show:

Theorem 1: A system with P processors and M memory implementing a bilinear program to compute the convolution of two N-1 degree polynomials must have average input bandwidth exceeding

$$B = \frac{I}{T} \geq \Omega \left(\frac{NP}{(M+P)(\frac{N}{M} + \log M)} \right)$$

where $P \leq M \leq N$.

The flexibility of this approach may not be apparent from the one example above. Essentially all that is needed is: to quantify the connectivity of the problem in terms of pebbling the inputs, and to find a reasonably efficient limited processor-limited memory implementation. For example, consider the problem of finding an algorithm independent bandwidth bound for Discrete Fourier Transform evaluation. (The FFT is one specific algorithm.) Using an approach similar to the Corollary of Lemma 1, Tompa (1978) has derived a lower bound on the inputs needed to evaluate any DFT using memory M even though the DFT does not contain a superconcentrator. (In fact it contains an N-hyperconcentrator, Valiant (1976).) Extending this result to the parallel pebble game with P processors we can show the following:

Theorem 2. Any P processor, M memory system using a linear algorithm to compute an N-point DFT must have bandwidth

$$B = \Omega \left(\frac{NP}{(M+P)(\frac{N}{M} + \log M)} \right)$$

where $P \leq M \leq N$.

Although we have only discussed bandwidth bounds for two related problems here, we are confident that bandwidth bounds for many problems can be derived using similar procedures because memory-time tradeoff bounds using the connectivity approach on several problems have already been described in the literature (see Tompa (1978) and Ja'Ja (1980)).

3. CONCLUSIONS

In this paper we have considered a technique for ameliorating pin restrictions. We have shown that an abstract model of computation, the pebble game, implies algorithm independent lower bounds for certain problems. (See Hong and Kung (1981) for an alternative model.) These bounds are tight to an order-of-magnitude over much of the design domain. Consider the extreme points for the DFT problem: at $M = P = N$, a full parallel FFT can be used to achieve bandwidth $B = 0$ (N/log N) matching Theorem 2; at $M = N$ and $P = 1$, the usual serial FFT gives $B = 0$ (1/logN) matching Theorem 2; but at minimum memory $M = \log N + 1$ and $P = 1$ the algorithm in Savage and Swamy (1978) gives $B = O(1)$ but Theorem 2 gives $B \geq \Omega(N/(N+\log N \log\log N))$.

The bound in Section 2 can be used to solve design problems such as the proper tradeoff between memory and processors to maximize the *area-to-bandwidth* ratio which is analogous to gate-to-pin ration observed in Section 1. Suppose that chip area is modelled by a function such as $A = M + a_p P^2$. The first term corresponds to memory area and the second terms models the processor area including interconnection area. With this or a similar model, Theorem 2 shows that to maximize the area-to-bandwidth ratio for a DFT system the memory area should be maximized. This leads to a lambda-bandwidth relation. On the other hand, if maximum usable memory, $M = N$, exists then any surplus area is used for processors resulting in a rho-bandwidth relation.

An alternative design objective such as minimizing computation time may lead to a quite different tradeoff. Given a fixed chip area and a model such as the one above, Lemma 3 illustrates a tradeoff between processor area and memory area. With this or a similar model, it can be shown that to minimize time, processor area should be proportional to memory area despite the pebble game effect which heavily penalizes systems with inadequate memory.

Finally, we emphasize the need for more pin conservation techniques. For example, bandwidth smoothing techniques should be used to reduce the instantaneous bandwidth (Section 1.1) to the average bandwidth. For example, we have developed a technique for smoothing the instantaneous input bandwidth of the FFT when $P = N$ from N to $N/\log N$ which is the average input bandwidth.

REFERENCES

DiGiacomo, J.J. (1980). "VLSI/LSI Circuit Functions: Their Challenges, Rewards and Problems", *IEEE Transactions on Components, Hybrids, and Manufacturing Technology, CHMT-3* 94-104.

Davis, C., Maley, G., Simmons, R., Stollar, H., Warren, R., and Wohr, T., (1980). "Gate Array Embodies System/370 Processor", *Electronics, Vol. 53, No. 22*, 140-143.

Ja'Ja J. (1980). "Time-Space Tradeoffs for Some Algebraic Problems", *12th Annual ACM Symposium on the Theory of Computing*, 339-350.

Kim, J. H., and Siewiorek, D.P., (1980). "Issues in IC Implementation of High Level, Abstract Designs", *Proceedings of the 17th Design Automation Conference*, 85-91.

Landman, B. S., and Russo, R. L., (1971). "On a Pin Versus Block Relationship for Partitions of Logic Graphs", *IEEE Transactions on Computers, C-20*, 1469-1479.

Pippenger, N., (1980). "Pebbling", *Proceedings of the 5th IBM Symposium on Mathematical Foundations of Computer Science*.

Russo, R. L., (1972). "On the Tradeoff Between Logic Performances and Circuit-to-pin Ratio for LSI", *IEEE Trans. on Computers, C-21*, 147-153.

Russo, P. M., (1980). "VLSI Impact on Microprocessor Evolution, Usage and System Design", *IEEE Journal of Solid-State Circuits, Sc-15*, 397-406.

Savage, J.E., and Swamy, S., (1978). "Space-Time Trade-Offs on the FFT Graph", *IEEE Trans. on Information Theory, IT-24*, 563, 568.

Tompa, M., (1978). "Time-Space Trade-Offs for Computing Functions, Using Connectivity Properties of Their Circuits", *10th Annual ACM Symposium on Theory of Computing*, 196-204.

Valiant, L. G., (1976). "Graph-Theoretic Properties in Computational Complexity", *Journal of Computer and System Sciences, 13*, 278-285.

Vuillemin, J., (1980), "A Combinatorial Limit to the Computing Power of V.L.S.I. Circuits", *21st Annual Symposium on Foundations of Computer Science*, 294-300.

OPTIMAL LAYOUTS FOR SMALL SHUFFLE-EXCHANGE GRAPHS

F. Thomson Leighton and Gary L. Miller

Mathematics Department
Massachusetts Institute of Technology
Cambridge, Massachusetts 02139, USA

1. INTRODUCTION

The shuffle-exchange graph is one of the best structures known for parallel computation. Among other things, it can be used to compute discrete Fourier transforms, multiply matrices, evaluate polynomials, perform permutations and sort lists [P80, S80, St71]. The algorithms needed for these operations are extremely simple and for the most part, require no more than logarithmic time and constant space per processor. The only exceptions are sorting lists (for which the best algorithm known requires $\Omega(\log^2 n)$ time) and performing arbitrary permutations (which may require $\Omega(\log n)$ space per processor).

With the development of integrated circuit technology, it has become possible to place large numbers of very simple processors on a single chip. Thus the question of how best to lay out the shuffle-exchange graph on a grid (using as little area as possible) has gained practical as well as theoretical importance. Thompson was the first to address the issue in the context of VLSI. In his thesis [T80], he showed that any layout of the n-node shuffle-exchange graph requires at least $\Omega(n^2/\log^2 n)$ area. In addition, he described a layout requiring only $O(n^2/\log^{1/2} n)$ area. Shortly thereafter, Hoey and Leiserson [HL80] improved the upper bound by finding an $O(n^2/\log n)$-area layout. By combining the techniques of Thompson, Hoey and Leiserson, both Rodeh and Steinberg [RS80] and Leighton, Lepley and Miller [LLM81] independently found $O(n^2/\log^{3/2} n)$-area layouts. The area layout question was finally settled by Kleitman, Leighton, Lepley and Miller [KLLM81] who employed entirely new methods to find an $O(n^2/\log^2 n)$ area-layout for the n-node shuffle-exchange graph, thus achieving Thompson's lower bound.

Although the $O(n^2/\log^2 n)$-area layout for the shuffle-exchange graph described in [KLLM81] is (up to a constant) *asymptotically* optimal, it is not optimal for small values of n (e.g., $n=2^7$). In fact, none of the general layout procedures thus far discovered [HL80, KLLM81, LLM81, RS80, T80] provide good layouts for small shuffle-exchange graphs. For practical applications, however, these are precisely the shuffle-exchange graphs for which we need good layouts.

In this paper, we describe techniques for finding good layouts for small shuffle-exchange graphs. Although the techniques do not yet constitute a general procedure for finding truly optimal layouts for all shuffle-exchange graphs, they can be used to find *very nice* layouts for *small* shuffle-exchange graphs. As examples, we have included layouts for $n = 8, 16, 32, 64$ and 128. The layouts are very nice in the sense that:
1) they require much less area than previously discovered layouts,
2) they have a certain natural structure which facilitates efficient layout description, chip manufacture and I/O management, and
3) they require the minimal amount of area for layouts with such structure.

2. PRELIMINARIES

(a) The shuffle-exchange graph

The *shuffle-exchange graph* consists of $n=2^k$ nodes and $3n/2$ edges. Each node is associated with a unique k-bit binary string $a_{k-1}\ldots a_0$. Two nodes ω and ω' are linked via a *shuffle edge* if ω' is a left or right cyclic shift of ω (i.e., if $\omega=a_{k-1}\ldots a_0$ and $\omega'=a_{k-2}\ldots a_0 a_{k-1}$ or $\omega'=a_0 a_{k-1}\ldots a_1$). Two nodes ω and ω' are linked via an *exchange edge* if ω and ω' differ only in the last bit (i.e., if $\omega=a_{k-1}\ldots a_1 0$ and $\omega'=a_{k-1}\ldots a_1 1$ or vice-versa).

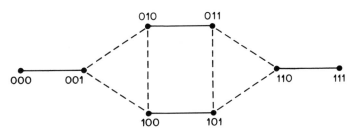

Fig. 1. The 8-node shuffle-exchange graph.

For example, we have drawn the *8*-node shuffle-exchange graph in *Figure 1*. In this figure, as well as throughout the rest of the paper, we have drawn the shuffle edges with *dashed lines* and the exchange edges with *solid lines*.

(b) Necklaces

The collection of all cyclic shifts of a node ω is called a *necklace* and is denoted by <ω>. For example, the necklace generated by *001* is <*001*> = {*001, 010, 100*}. Note that each necklace corresponds to a cycle in the shuffle-exchange graph (see *Figure 1*) and that shuffle edges always link nodes which are in the same necklace.

If a necklace contains precisely k nodes, then it is said to be *full*. Otherwise a necklace contains less than k nodes and is said to be *degenerate*. For example, <*001*> is full while <*000*> is degenerate.

The partition of the shuffle edges into necklaces is a key part of the layout technique described in *Section 3*.

(c) The Thompson model

In what follows, we will descibed layouts for the shuffle-exchange graph in terms of the *grid model* developed by Thompson [T80]. In this model, processors are represented by points which are located at the intersection of grid lines. Wires connect pairs of processors and must follow along grid lines. Two wires can cross each other but only at the intersection of grid lines (i.e., two wires cannot overlap for any distance). In addition, wires are not allowed to overlap processors. The *area* of a layout is defined to be the product of the number of vertical tracks and the number of horizontal tracks containing a processor or wire of the network.

As an example, we have included a Thompson model layout of the *8*-node shuffle-exchange graph in *Figure 2*. This layout requires *2* horizontal tracks and *6* vertical tracks, thus having area *12* (which is optimal). For simplicity, we have replaced the *3*-bit binary string associated with each node by its *numeric value*.

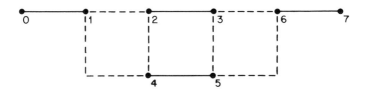

Fig. 2. An optimal Thompson model layout for the 8-node shuffle-exchange graph.

(d) Rational for using the Thompson model

We have chosen to use the Thompson model to illustrate our techniques because of its widespread acceptance and its simplicity. Although the assumption that processors can be represented by points is clearly false in practice, good Thompson model layouts can still be used to develop good practical layouts. The manner in which a Thompson model is useful varies with the size of the processors involved. For example, if one desires to use the shuffle-exchange graph as a permuter, then each processor need only contain k storage registers and some I/O hardware. Such a processor can easily be hardwired in a k by k square. In order to achieve maximum parallelism, each wire of the Thompson model layout is reproduced k times so that an entire k-bit word can be transmitted in each time step. For example, the optimal Thompson model layout in *Figure 2* (where $k=3$) can be transformed into the more realistic $6x18$ layout shown in *Figure 3* by tripling the grid lines and replacing point processors by $3x3$ boxes (into which the guts of each processor will later be wired).

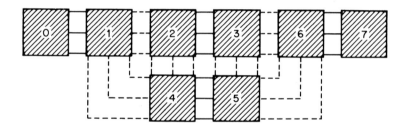

Fig. 3. A transformed Thompson model layout of the 8-node shuffle-exchange graph.

For some applications, the processors themselves require an entire chip. For example, every processor of a shuffle-exchange graph used to compute discrete Fourier transforms must be equiped with a floating point multiplier. Using the best technology currently available, only a few floating point multipliers can be wired onto a single chip. In this case, a Thompson model layout can be used to design an efficient *layout of chips* where each chip contains a single processor. (Such a device is currently under development at IBM). The wires, as before, are replicated to achieve maximum parallelism but now serve as links between chips. Since the wires must be much wider in such a device, the side length of a processor (the chip) is about the same as the combined width of all the wires (pins) attached to it. By following an expansion procedure similar to the one described in the previous example, a good Thompson model layout can thus be used to design a good practical layout.

3. LAYOUT TECHNIQUES

(a) A general class of layouts

In what follows, we will consider layouts of the shuffle-exchange graph for which:

1) each necklace appears as a rectangle consisting of arbitrarily long segments of two vertical tracks and unit length segments of two horizontal tracks,

2) the horizontal tracks are divided into pairs, each pair containing at most one full necklace and any number of degenerate necklaces, and

3) each exchange edge appears as a horizontal line segment.

For example, the layouts in *Figures 2* and *4-7* have this form. In [LLM81], Leighton, Lepley and Miller show that such a layout exists for every shuffle-exchange graph. In fact, by rearranging the left-to-right ordering of the necklaces and/or the top-to-bottom ordering of the exchange edges, it is easy to produce a large class of such layouts for any shuffle-exchange graph. In particular, we will be interested in those layouts which use the smallest number of vertical and horizontal tracks. *The layouts in Figures 2 and 4-7 are optimal in this respect.*

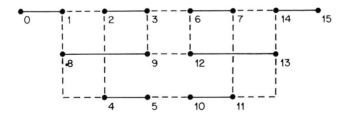

Fig. 4. A 3x8 Thompson model layout for the 16-node shuffle-exchange graph.

As is easily observed, these layouts require a surprisingly small amount of area. Further, the structure of the layouts facilitates efficient description, chip manufacture and data management. For example, for small values of n, it is feasible to attach a pin to each of the $\Theta(n/\log n)$ necklaces, thus allowing n values to be loaded into an n-node shuffle-exchange graph in just $O(\log n)$ steps.

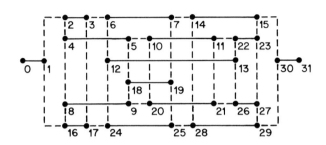

Fig. 5. A 6x14 Thompson model layout for the 32-node shuffle-exchange graph.

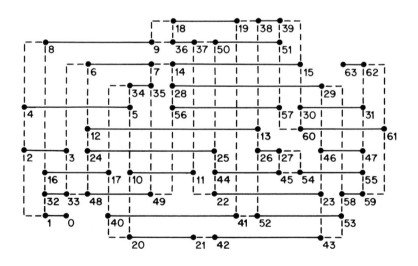

Fig. 6. An 11x18 Thompson model layout for the 64-node shuffle-exchange graph.

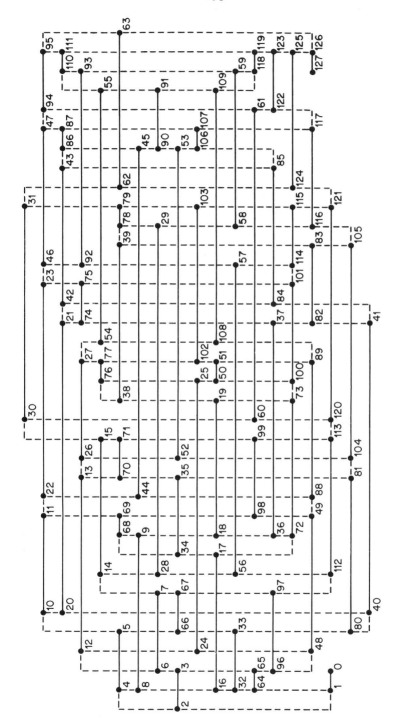

Fig. 7. A 19x36 Thompson model layout for the 128-node shuffle-exchange graph.

(b) Optimization Techniques

In order to find the layouts in *Figures 2* and *4-7*, we employed a combination of heuristics and exhaustive searches. The basic techniques are summarized in the following three steps.

Step 1: Partition the shuffle edges into necklaces. Order the necklaces linearly from left to right so that the number of exchange edges that overlap at each point of the ordering is kept small. More precisely, find an ordering of the necklaces for which the maximum number of exchange edges overlapping at any point is minimized. For example, no more than *6* exchange edges overlap at any point of the ordering used to produce the layout for the *32*-node shuffle-exchange graph shown in *Figure 5*. If we switch the necklace <*5*> with <*11*>, then *9* exchange edges would overlap in the gap between <*7*> and <*5*>. Since the maximum overlap is a lower bound on the number of horizontal tracks necessary to insert the exchange edges, we can easily see that the latter ordering is inferior since any layout it produces must have at least *9* horizontal tracks. Note that the layout in *Figure 5* has just *6* horizontal tracks.

It is not known how best to order the necklaces in general. For small shuffle-exchange graphs, however, there are several simple heuristics which produce optimal orderings. For example, if we initially order the necklaces from left to right so that nodes with binary numbers containing r zeros are placed to the left of nodes containing $r-1$ zeros for $1 \leq r \leq k$, then an optimal ordering can be produced by making at most one or two switches. The optimal ordering for $k=5$ ($n=32$) shown in *Figure 5* was produced by this method. Note that no switches were needed in this case. (See [LLM81] for a detailed theoretical discussion of such layouts).

A different but equally useful heuristic orders the necklaces from left to right so that the minimum value of the nodes in each necklace form an increasing sequence. The ordering in *Figure 5* could also be produced by this method. In this case, the minimal values are *0, 1, 3, 5, 7, 11, 15* and *31*. Layouts produced by this method are discussed from an asymptotic point of view in [KLLM81].

Probably the most difficult task is proving that a good ordering is optimal. The techniques we have used to prove optimality depend heavily on exhaustive searches. For $k \leq 8$, the techniques have succeeded in proving the optimality of good orderings. For $9 \leq k \leq 13$, we have found good orderings but have not been able to determine whether or not they are optimal. We have summarized the results in *Table 1*. Note that for each k, the maximum overlap of the best known ordering serves only as a lower bound on the number of horizontal tracks that will be required for any layout with that ordering. In some cases (e.g., $k=6,7$), additional horizontal tracks may be required.

TABLE 1

Maximum Overlap of Best Known Ordering

k	n	maximum overlap	optimal?
3	8	2	yes
4	16	3	yes
5	32	6	yes
6	64	10	yes
7	128	18	yes
8	256	33	yes
9	512	62	?
10	1024	115	?
11	2048	214	?
12	4096	388	?
13	8192	754	?

Step 2: The next step is to insert the exchange edges using as few horizontal tracks as possible. The techniques we used to find the layouts in *Figures* 2 and 4-7 are extensions of the theoretical ideas developed in [LLM81]. In that paper, the authors use the structural properties of the shuffle-exchange graph to produce layouts of the desired form but with an excessive number of horizontal tracks.

In our layouts, we first insert those exchange edges which cross a region of maximum overlap. To do this, we (for the most part) follow the top-to-bottom ordering given in [LLM81], making sure that every (or, as in the case with $k=6$ and 7, nearly every) horizontal track contains an edge in the region of maximum overlap. We next insert the exchange edges which cross neighboring regions in such a way that no new horizontal tracks are required. This process continues until all of the exchange edges are inserted. If done carefully, the number of horizontal tracks used will be the same as or only slightly larger than the maximum overlap.

Step 3: The third and final step is to adjust (if possible) the exchange edges so that degenerate necklaces can be doubled up with full necklaces and inserted into the same pair of vertical tracks. Since degenerate necklaces have substantially fewer nodes than do full necklaces, it is usually possible to accomplish this task without increasing the number of horizontal tracks used. For example, the number of vertical tracks needed to lay out the *16*-node and *64*-node shuffle-exchange graphs can be substantially reduced by this procedure.

(c) Other layouts

To this point, we have considered only a specific class of layouts for the shuffle-exchange graph. As these layouts are quite good, it is not clear that we need to consider others. Nevertheless, it is worth pointing out that slightly better layouts do exist for some shuffle-exchange graphs. For example, by considering layouts in which the exchange edges are allowed to bend and in which two or more full necklaces can occupy the same pair of vertical tracks, it is possible to construct the layout for the *32*-node shuffle-exchange graph shown in *Figure 8*.

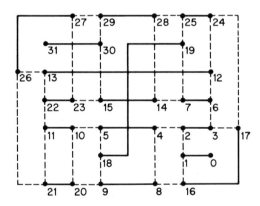

Fig. 8. A 7x9 Thompson model layout for the 32-node shuffle-exchange graph.

It is likely that slight improvements can also be made for larger shuffle-exchange graphs. At this point, however, we feel that research efforts should be directed more towards implementation of the good layouts already discovered. Once this is done, it will be much clearer whether or not the effort necessary to further reduce the layout area is justified.

4. ACKNOWLEDGEMENTS

This research was supported in part by the Advanced Research Projects Agency of the Department of Defense and monitored by the Office of Naval Research under contract number N00014-C-80-0622.

In addition, we would like to thank the following people for their helpful remarks and suggestions: D. Kleitman, C. Leiserson, M. Lepley, R. Rivest and R. Zippel.

5. REFERENCES

[HL80] D. Hoey and C.E. Leiserson, "A Layout for the Shuffle-Exchange Network," *Proceedings of the 1980 Conference on Parallel Processing.*

[KLLM81] D. Kleitman, F.T. Leighton, M. Lepley and G.L. Miller, "New Layouts for the Shuffle-Exchange Graph," *Proceedings of the 13th Annual ACM Symposium on the Theory of Computing,* Milwaukee, Wisconsin, 1981.

[LLM81] F.T. Leighton, M. Lepley and G.L. Miller, "New Layouts for the Shuffle-Exchange Graph Based on the Complex Plane Diagram," to appear as an MIT Tech Report.

[L80] C.E. Leiserson, "Area-Efficient Graph Layouts (for VLSI)," *Proceedings of the 21st Annual Symposium on Foundations of Computer Science,* Syracuse, New York, 1980.

[P80] D.S. Parker, "Notes on Shuffle-Exchange Type Switching Networks," *IEEE Transactions on Computers, C-29, 3 (March, 1980),* pp. 213-222.

[RS80] M. Rodeh and D. Steinberg, "A Layout for the Shuffle-Exchange Graph with $O(n^2/log^{3/2}n)$ Area," Technical Report 088, IBM Israel Scientific Center, 1980.

[S80] J.P. Schwartz, "Ultracomputers," *ACM Transactions on Programming Languages and Systems, Vol. 2, No. 4 (October, 1980),* pp. 484-521.

[St71] H.S. Stone, "Parallel Processing with the Perfect Shuffle," *IEEE Transactions on Computers, C-20, 2 (February, 1971),* pp. 153-161.

[T80] C.B. Thompson, *"A Complexity Theory for VLSI,"* Ph.D. Thesis, Carnegie-Mellon University Computer Science Department 1980.

WIRING SPACE ESTIMATION FOR RECTANGULAR GATE ARRAYS

W.E. Donath* and W.F. Mikhail°

*IBM Thomas J. Watson Research Center
Yorktown Heights, New York 10598

°IBM General Technology Division, East Fishkill
Hopewell Junction, New York 12533

INTRODUCTION

Wirability estimation is of crucial importance in VLSI chip design, since wiring space usually dominates circuit space requirements. In the absence of a reasonable estimate of wiring space requirements, excessive effort would be required for physical design, and a designer might not realize that the circuitry does not fit on the given chip area until too late in the design cycle.
In previous work, Donath (1978) and Heller, et. al., (1978) presented practical methods for estimating average wire length of interconnections in square arrays and a stochastic model for using the wire length estimate to derive actual wiring space requirements. This prior work made extensive use of the observed partitioning relationship called Rent's Rule, as discussed by Landman and Russo (1978); Donath (1974), whereby the average terminal count T is related to the block count C and two constants A and p as:

$$T = A \cdot C^p \qquad (1)$$

In the stochastic wirability theory Heller, et. al., (1978) developed a one dimensional model; in the two dimensional extension of this model, we divide the wiring area into horizontal and vertical channels, where the width of the channels is heuristically set to correspond to what one might call the 'mobility' of the wire segments (i.e., how far to the right or to the left a typical vertical segment might be moved by the wiring algorithm). The width of a horizontal (vertical) channel is related to the average vertical (horizontal) length a wire with non-zero horizontal component has. For the square case we assume symmetry. In this paper we extend the method to estimate the allocation of horizontal versus vertical tracks.

This paper is organized as follows: first, we treat the mathematical model and derivations, than we give some experimental results, and finally we discuss our conclusion.

MATHEMATICAL MODEL

We consider the rectangular array to be made up of a concatenation of square arrays (as in fig. 1): the first part of this section considers the classification of intra- and inter- square connections; the second part deals with the wire lengths, and the third part treats the input parameters to the stochastic model.

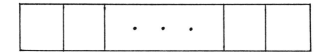

Fig. 1 *Concatenation of square arrays.*

Our input parameters to this problem are:
  Y = number of wires (connections) per gate
  p = the Rent exponent of equation (1)
  m = the number of gates per square array
  C = the number of gates in the chip

The average number of terminals of a complex of k gates is given by equation (1). Let γ be the average number of connections per terminal. It is to be observed that γ is a function of the number of gates and thus it varies as we partition the rectangular array into square arrays. However, we shall assume γ to be constant since we believe the variation to be small. Hence, the average number of wires (connections) external to a block of k gates is given by:

$$W_k = A \, \gamma \, k^p \qquad (2)$$

Now, we derive the fraction f of wires that interconnect between the square arrays. From equation (2), the fraction of wires external to an array of C gates is given by:

$$F_C^E = A\gamma C^P/A\gamma C = C^{P-1}$$

and thus the fraction of wires internal to the array is:

$$F_C^I = 1 - F_C^E = 1 - C^{P-1} \quad (3)$$

Also, the fraction of wires external to the square arrays (but still internal to the rectangular array of C gates) is given by:

$$F_E = m^{P-1} - C^{P-1} \quad (4)$$

From equations (3) and (4), it follows that f is given by:

$$f = \frac{F_E}{F_C^I} = (m^{P-1} - C^{P-1})/(1 - C^{P-1}) \quad (5)$$

Therefore, (1-f) is the fraction of wires internal to the squares in fig. (1).

There are four sets of wire segments – horizontal and vertical segments of the wires internal to the squares, the horizontal wires interconnecting the squares, and the vertical segments, which line up the endpoints of the wires connecting the various squares. For each set of segments we compute Y, the number of wires per gate, R1, the average length in the same direction, and R2, the average length of segments connected to those wires perpendicular to their direction.

The average length R^S of the wires inside the squares, expressed in terms of gate spacing, is given by Donath (1979) as:

$$R^S = .6 \; \frac{2}{9} \left(\frac{m^{P-1/2} - 1}{4^{P-1/2} - 1} - \frac{1 - m^{P-3/2}}{1 - 4^{P-3/2}} \right) \quad (6)$$

Let g be the fraction of wires with only one component (either horizontal or vertical); then we use the derivation given by Heller, et. al., (1978) to find number of components per gate, Y, as well as their R1 and R2 values as:

$$Y^S = Y(1-f)(1-g/2)$$

$$R1^S = R^S/(2-g) \quad (7)$$

$$R2^S = (R^S - 1)/(2-g)$$

where the superscript s stands for wires inside the square. For the horizontal components of the wires interconnecting the squares we use:

$$R^{bx} = h \; m^{1/2} \; \frac{5}{3} \; \frac{(1-4^{p-1})}{(1-(C/m)^{p-1})} \; \frac{((C/m)^p - 1)}{(4^p - 1)} \quad (8)$$

where h is an adjustment factor, which we found to give us the best results when equal to 0.85. For vertical components of these wires we use:

$$R^{by} = h \; m^{1/2}/3 \quad (9)$$

The superscript b stands for between square wires. Noting that the fraction of wires crossing between the square arrays is f and that each inter-connection has an x component, we now find for the Y, R1, and R2 values of the between the squares wires the following:

$$Y^{bx} = Y \; f$$
$$R1^{bx} = R^{bx} \quad (10)$$
$$R2^{bx} = R^{by}$$

and since some of the inter-squares wires have only x segments and no y segments, we obtain:

$$Y^{by} = Y \; f \; (1 - 1/(h \; m^{1/2}))$$
$$R1^{by} = R^{by} \quad (11)$$
$$R2^{by} = R^{bx}$$

We denote by superscripts x and y averages over all wires in the x and y direction; we find

$$Y^x = Y^s + Y^{bx}$$
$$R1^x = (Y^s R1^s + Y^{bx} R1^{bx})/Y^x \quad (12)$$
$$R2^x = (Y^s R2^s + Y^{bx} R2^{bx})/Y^x$$

and similarly for the y direction. The effective input parameters for the stochastic model in the x direction as given by Heller, et al. (1978) are:

$$Y^x_{eff} = Y^x(1+R2^x)$$

$$C^x_{eff} = C/(1+R2^x) \qquad (13)$$

$$R^x_{eff} = R1^x$$

The output parameter is the number of horizontal tracks per gate, T^x, that corresponds to a given probability of successful wiring. Following the procedure outlined by Heller, et al. (1978)

$$T^x = T^x_{eff}/(1+R^x_2) \qquad (14)$$

Similar formulas for the input and output parameters in the y direction can be obtained by replacing the superscript x by y in equations (13) and (14).

EXPERIMENTAL RESULTS

A study was performed using a chip, which had 1326 internal connections and 581 gates, and which was studied in 7 types of arrays - (10x68), (14x49), (18x38), (26x26), (38x18), (49x14) and (68x10). Placement was done for each of those configurations using the placement procedure given by Khokhani and Patel (1977). These chips were wired using a wiring procedure described by Chen, et. al., (1977). The number of horizontal and vertical tracks per cell was varied in such a way that of the order of 15-25 overflows were observed, which we suspect can easily be embedded. We selected the pair of values T^x, T^y in such a way that their sum is a constant, but among the possibilities of allocating this quantity we selected the one which yielded the least number of overflows.

In Table I, the placement results for the average connection length measured in cell to cell pitches, are compared with the theoretically calculated values. It can be seen that experiment and theoretical value are in satisfactory agreement. The wiring results are given in Table II. Here, notice that while the total track values are in satisfactory agreement with the theoretically computed values, the allocation of horizontal and vertical tracks differs; the reason for that is that access to the gates is mostly through the vertical tracks, which are therefore used more than the horizontal tracks. It is interesting to notice that by removing one track from the horizontal direction and adding it to the vertical direction, reasonable agreement can then be obtained between the adjusted values and the empirically observed values.

We also notice from Table II, in the experimental columns that while the number of overflow wires has a shallow minimum, it increases sharply as we deviate significantly from the horizontal-vertical values that yield the minimum; i.e., if we remove 2 or more tracks from one direction to another. This demonstrates the importance of proper track allocation.

In Table III, we give array areas and aspect ratios for the various array configurations ignoring the active area of the gates. The area dimensions are expressed in track counts and the aspect ratio is defined as the larger ratio of the total number of tracks per array in one direction (horizontal or vertical), to that in the other direction. From the table, we observe the following:

    a. For all array configurations, other than the first, there is a reasonable agreement between the theoretical and experimental results for the area.

    b. The observed number of overflow wires for the (10x68) array is almost half that of the other arrays. This implies that we require a smaller number of tracks per gate than twenty. We conducted an experiment using 19 tracks allocated horizontally and vertically as (12,7); (11,8) and (10,9). The observed numbers of overflows were 46, 43 and 66 respectively. This suggests that probably 19.5 tracks would have been a better choice, in which case the area comparison would be satisfactory.

    c. The aspect ratio is a sensitive parameter e.g., in the (14x49) array configurations the (9,9) and (10,8) track allocations almost yield the same number of overflows and a selection of the latter track allocation results in an aspect ratio change from 3.5 to 2.8 which is closer to the theoretical value. Similar results can be obtained for the (49x14) array configuration.

    d. If one deviates from equal numbers of rows and columns per array, it appears that some benefit in array area would be obtained if the number of rows exceeds the number of columns.

CONCLUSIONS

A method has been given to estimate wire lengths and track requirements for rectangular gate arrays, whereby one can allocate the tracks in the horizontal and vertical directions. Reasonable agreement between theory and one experimental test case has been observed. The results obtained in this paper can be used by the gate array designer to determine the optimum ratio of rows to columns in the array according to his cell layout and the final shape of the chip. Also they can be used in custom chip layout to determine the area of a subunit as a function of its aspect ratio.

ACKNOWLEDGEMENTS

The authors gratefully acknowledge and appreciate the help provided by E. Schanzenbach in the implementation of the experiments. They would like to express their thanks to Dr. C.G. Hsi and Dr. J. Lee for their valuable comments and suggestions during the course of this study. Dr. M. Feuer and Dr. R. Russo went through the manuscript and made many recommendations, both in the substance and style of the paper, for which the authors are grateful.

REFERENCES

Chen, K.A., Feuer, M., Khokhani, K.H., Nan, N., Schmidt, S. (1977). The Chip Layout Problem: An Automatic Wiring Procedure, proceedings of 14th Design Automation Conference, pp. 298-302.
Donath, W.E. (1974). Equivalence of Memory to Random Logic, in IBM Journal of Res. and Dev., vol. 18, pp. 401-407.
Donath, W.E. (1979). Placement and Average Interconnections Length of Computer Logic, in IEEE Trans. on Circuits and Systems CAS-26, pp. 272-276.
Heller, W.R., Mikhail, W.F. and Donath, W.E., (1978). Prediction of Wiring Space Requirement for LSI, in Journal of Design Auomation and Fault Tolerant Computing, pp. 117-144.
Khokhani, K.H. and Patel, A.M. (1977). The Chip Layout Problem: A Placement Procedure for LSI, proceedings of 14th Design Automation Conference, pp. 291-297.
Landman, B.S. and Russo, R.L. (1978). On a Pin Versus Block Relationship for Partitions of Logic Graps, IEEE. Trans. on Computers, C-20, pp. 1469-1471.

TABLE 1

Average Connection Length
(measured in gate pitches)

Array Configuration (rows x columns)	Experimental Results	Theoretical Results
10 x 68	4.39	4.0
14 x 49	3.67	3.41
18 x 38	3.37	3.16
26 x 26	3.07	2.87
38 x 18	3.19	3.16
49 x 14	3.44	3.41
68 x 10	3.99	4.0

TABLE II

Wiring Results

Experimental					Theoretical (unadjusted)			Adjusted	
Array Confg.	Total # Trks/ Gate	T^x	T^y	Obsrvd # of Overflws	Total # Trks/ Gate	T^x	T^y	T^x	T^y
10x68	20	14	6	49	19.6	13.9	5.7	12.9	6.7
		13	7	25					
		12	8	12					
		11	9	27					
		10	10	47					
14x49	18	10	8	25	17.8	11.1	6.7	10.1	7.7
		9	9	24					
		8	10	40					
18x38	17	9	8	27	17.1	9.7	7.4	8.7	8.4
		8	9	20					
		7	10	32					
26x26	16	9	7	63	16.4	8.2	8.2	7.2	9.2
		8	8	27					
		7	9	24					
		6	10	32					
		5	11	58					
38x18	17	7	10	27	17.1	7.4	9.7	6.4	10.7
		6	11	25					
		5	12	39					
49x14	18	6	12	34	17.8	6.7	11.1	5.7	12.1
		5	13	32					
		4	14	39					
68x10		7	12	72	19.6	5.7	13.9	4.7	14.9
		6	13	28					
		5	14	22					
		4	15	63					
		3	16	133					

TABLE III

Areas and Aspect Ratios of Arrays

Array Configuration	Experimental Results		Theoretical Results	
	Area	Aspect Ratio	Area	Aspect Ratio
10x68	65.3K	4.53	58.8K	3.5
14x49	55.6	3.5	53.4	2.7
18x38	49.2	2.38	50	2.03
26x26	42.6	1.28	44.8	1.28
38x18	45.1	1.15	46.8	1.26
49x14	44.6	1.35	47.3	1.65
68x10	47.6	2.4	47.6	2.14

SESSION 7

IMPACT OF TECHNOLOGY ON THE DEVELOPMENT OF VLSI

M.W. Larkin

*Managing Director, Plessey Solid State Division
Swindon SN2 6BA, UK*

INTRODUCTION

The concept of the Integrated Circuit goes back to the early days of the vacuum tube or valve, when some compound valves were made. However, the first solid state integrated circuit was brought to practical realization in 1959. We can therefore now review the progress which has been made over the past 20 years, which has brought us from discrete semi-conductor devices to VLSI components containing up to 100,000 active elements. The technological innovations which have been made in silicon processing during that period and the general trends in the industry which have influenced the manufacture of integrated circuits, can be considered; particularly the influence which processing capability has on the size and performance of circuits which can be made. Based on this 20-year history, it is interesting now to try to project a course of future development of semiconductor technology as it is applied to VLSI, particularly with regard to the problems which arise in evaluation of the VLSI circuits themselves.

The concept of the Integrated Circuit goes back to the early days of vacuum tube or valve technology when some compound valves were made and passive components were included inside the same vacuum envelope. However, the first proposal for a solid state integrated circuit based on semiconducting material was made in 1955 by G.W.A.Dummer at the Royal Research Establishment, Malvern. This concept was brought to practical realization in 1959 simultaneously by Westinghouse Electric Corporation and Texas Instruments, under the US Air Force Molecular Electronics Programme. However, at that time there were severe limitations to the degree of integration which could be achieved due to the scope of the available technology. This was still based on shaping of the semiconductor substrate to define junction area, usually by selective chemical etching, or, indeed, in some cases by mechanical milling of the semiconductor, using the MESA technology. The metallising patterns had to be defined by means of selective deposition of evaporated aluminium through a metal mask interposed between the semiconductor and the evaporating source. This meant that the best resolution of material dimension which could be achieved at that time was approximately 25 microns, and the usual line width was of the order of 100 microns. It was not until the introduction of planar technology which utilised the silicon oxide as a mask for defining diffused areas in 1961, that the full potential of integration of a single piece of silicon could be realised. By combining the use of selectively etching silicon oxide to define diffused junction areas, together with selectively etched metallised layers to define interconnect patterns, it was possible not only to make a fully integrated circuit with passivated junctions, but also to align secondary patterns building up the structure with increased precision, thereby using the techniques of colour printing on a reduced scale, to create a fully integrated circuit in silicon. Once it had been realised that techniques of microlithography could not only be applied for defining area with precision, but could also be used to replicate the patterns in a uniform manner, the path was open towards mass production of integrated circuits.

In 1961, planar circuits used line widths of 25 microns and had a typical packing density of 4 gates of a die size of 18 sq.mm. These early circuits sold for approximately £1.50p each. In 1981, planar circuits such as the 16000-bit Random Access Memory, use line widths of 3 microns and have the equivalent of 35000 gates on a die size of 17.3 sq.mm. They also sell for approximately £1.50p each.

Let us now trace the reasons for this dramatic size reduction and cost improvement through the years. The introduction of photolithographic techniques gave immediate access to reducing the size of integrated circuits by proving the resolution of the pattern definition on the photographic masks and initially this was regarded as a means for increasing the content of a single integrated circuit for the purpose of weight and size reduction. It soon became apparent, however, that the manufacturing yield of integrated circuit die was strongly related to the active area of the device. This is due to the effect of random defects both in the base silicon material, and those introduced during each stage of processing. Therefore, major improvements in yield could be achieved by the reduction in the magnitude and size of the defects and also by reducing the active area of the circuit itself.

The actual cost of the circuit with regard to silicon area has not effectively changed, and if we consider the end sale value per sq.mm. of silicon across the complete spectrum of discrete semiconductor devices up to VLSI, it can be seen that, effectively, this is relatively constant independent of the size of the device. Therefore, one of the major influences of technology on VLSI has been the increase in complexity of the circuit content on a fixed area, which has gone up by 4 orders of magnitude over the past 20 years. This has generated a dramatic decrease in the cost per function of the electronics circuitry. This achievement on its own would not have led to the rapid increase in the use of integrated circuits. However, a parallel development was taking place in the design of electronics systems. It was becoming apparent that the approach of using a computer programmed by means of software could simulate functions of hard wired circuitry. However, it was not until low cost electronics in the form of integrated circuits were available that the full potential of this microprocessor approach could be exploited.

In the 1950's, a typical computer was 16ft.long, 8ft.high and 4ft.wide, and it contained 4,000 valves, 6 miles of wire and 100,000 soldered joints. It cost more than £200,000 and needed 27kw of power to function. The equivalent microprocessor computer today costs less than £20, is 1/4" square by .015" thick, and operates on 5 milliwatts. It is millions of times more reliable than the early available equipment which would function continuously for less than 30 minutes.

If we now attempt the extrapolation of this advance over the next 20 years, it is difficult from our current standpoint to forecast advances of this magnitude. Indeed, it is extremely difficult to forecast what kind of advance will be made in microelectronics within the next 5 years! The best that can be done is to review recent developments in processing technology and consider some of the current limitations of their development and exploitation.

Since the initial introduction of planar technology, there have been several significant silicon processing techniques which have been implemented to improve the fabrication of planar device structures. Epitaxial growth of doped silicon layers was used to create defined layers and achieve well-defined junctions for isolation, which could not be fabricated as readily by diffusion. The control of the introduction of impurities has also been greatly improved by the introduction of ion implantation techniques. The implantation of impurities into silicon by means of an ion beam has given another tool for junction formation which can be used to selectively define doped areas either by localised deposition or by the use of oxide masking. Both of these techniques lead to improved processing control and hence higher yield. A more recent innovation which is still under development, is the substitution of plasma etching in place of the use of liquid chemicals. This, also, reduces the probability of the introduction of unwanted impurities during processing.

The resolution of the photographic patterns has been continuously improved by - first, the introduction of improved conventional optical equipment, - then, the convertion to projection alignment, - and, finally, by the fabrication of masks using electronic beam exposure. It is now possible, in a routine manner, to produce 3-micron feature sizes, and in future, 1-micron feature sizes will be commonplace. However, these advances have not been without some difficulty. The use of direct electron beam exposure on to the silicon wafer has not been universally applied, since the time required to define the high resolution areas is too long for mass production techniques. In order to apply direct electron beam fabrication, it will be necessary to develop either a higher sensitivity resist material, or some means of simultaneous electron beam writing on to the silicon substrate.

A further major advance in the development of complex LSI circuits was the development of the MOS transistor structure, which is much simpler to fabricate than the equivalent bipolar device. Due to the simplified structure and the reduced number of processing steps, MOS circuits can be made with a higher packing density and a higher yield than bipolar circuits. It is for this reason that the majority of VLSI circuits are based on MOS devices.

As we proceed to more and more demanding parameters in wafer fabrication, the equipment required is even more sophisticated and expensive. The cost of this equipment has to be offset against a product whose selling price is decreasing year by year. In order to recover the capital investment it is necessary that this equipment should be used continuously at a high volume throughput. The industry is becoming more and more orientated towards mass production. The dilemma we must now face is that, as we go towards higher levels of integration, the number of units required for each particular application will go down and therefore we must introduce some flexible method of fabrication which allows high volume production methods to be linked into unique customer requirements. These requirements, in part, are being met by the concept of flexible interconnection which was first proposed by Macintosh & Green in 1964. Many semiconductor companies are now introducing the concept of logic arrays or modular cell approaches for custom design. In order to exploit this capability it is necessary that a very sophisticated capability for computer aided design is generated. It is also necessary that the integrated circuit design must be related to the system need at a very early stage of the design development.

The past 5 years have shown an increase in application of microprocessor control to wafer fabrication in order to improve the consistency and control of process parameters. We have now reached a time interval where process technology is awaiting the development of design techniques to fully exploit present capabilities of wafer manufacture. If we look beyond the 1980's, it is certainly feasible that process capabilities could be extrapolated to the sub micron level. However, other major questions arise regarding the application of the products which we could then manufacture, and the test and verification of their function. If we consider sub micron devices, we are contemplating a volume capability for the manufacture of a large computer, e.g. an IBM 370, on a single silicon die, less than 30 sq.mm. At this size, based on our present experience, the yield would be sufficiently good that such devices could be produced in volume for less than £1.50p.

But let us consider the problem of testing such a product. How would it be fully evaluated? Should we consider in-built functional redundancy? It is difficult enough to test our current large-scale integrated circuits at the level of a 64K RAM, or the equivalent logic element. How would we fully evaluate the computer on a chip? What do we do if it does not function correctly? Would we be able to trace the defect and correct it? It is relatively simple in a large system to partition the system into discrete blocks and isolate defects in this manner, but when the system is fully integrated, this is no longer possible. Electron beam testing of such products has been tried in the past but the results of such analysis are, at the best, ambiguous. However, it may be that the improvement of such a technique, using a fast scanning beam to evaluate sub micron areas, may be the one way in which circuits of this kind will be able to be tested in a reasonable time.

I therefore consider that this conference faces a challenge. We have reached a phase in the development of integrated circuit technology where the processing capability is able to manufacture circuits at a level beyond our capability to design and beyond our current capability of test - perhaps even beyond our present capability to use. It is up to the systems engineers to become involved at the design level in order to ensure that we can fully exploit our current capability.

CONCLUSIONS

Over the past 20 years we have seen a dramatic change in the level of integration which can be achieved on a single piece of silicon, primarily due to the improved control which has been exercised over the processing technology. We have now reached a stage where the technology itself is capable of yielding even further improvement in packing density of the circuitry. However, we may be facing a limitation in our capability of evaluating the circuits themselves in order to guarantee their performance. It is possible that future developments in VLSI may be more constrained by computer aided design and testing, than they are by the physical limitations of processing.

A GENERAL CELL APPROACH FOR SPECIAL
PURPOSE VLSI-CHIPS

K.D. Mueller-Glaser and L. Lerach

Data Processing Department and Components Department
SIEMENS AG, D-8000 Munich, F.R. Germany

1. INTRODUCTION

Some of the main goals for the manufacturer of digital systems are reducing costs, increasing realibility, decreasing power consumption and system volume, enhancing performance and last but not least conserving compatibility. The first goals force the application of large and very large scale integrated circuits. Standard LSI and VLSI circuits are used as far as possible especially for new designs of system parts. However the integration of whole systems with all their variety in digital or analog interfaces requires the design of customized VLSI chips. Other reasons originate in the restriction to conserve system compatibility because writing or rewriting software to adapt standard VLSI Processors could be too costly and emulating a system built
with standard TTL or ECL circuits by the use of standard VLSI circuits mostly decreases system performance. Therefore in the future most of the designed VLSI-chips will be customized.

However a serious problem occurs with the integration of a whole system or a part of a system into a VLSI-chip. On the one hand the design of the chips is extremely expensive. On the other hand the number of chips required approaches one per system which means only several thousands of pieces a year. Design philosophy for customized VLSI chips therefore is different to that for standard VLSI chips. Chip area and resulting yield is no longer the first goal but design cost, design time, design security and design for testability are the goals.

The SIEMENS Data Processing Department in cooperation with the Components Department and the Central Research and CAD-Lab's has developed a methodology for the economic design of customized VLSI-chips containing 5000 to 20000 gate functions plus on chip RAM or ROM and with relatively low production volumes of several thousands of pieces a year. Design cost and design time are reduced by using a general cell approach and an integrated suite of CAD programs. The present technology used is a 3 um N-Channel, Silicon-Gate, Depletion-Load-MOS.

2. A GENERAL CELL APPROACH FOR SPECIAL PURPOSE VLSI-CHIPS

Figure 1 shows the general trade-off between design time and chip area of several design philosophies currently in discussion (Mudge, 1980).

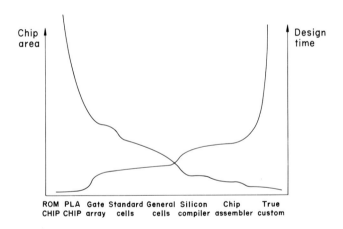

Fig.1 Trends of design time versus chip area for several design philosophies.

For VLSI-Systems pure ROM or PLA chips are not feasible because of large chip area and high access times. On the other hand fully optimized custom chips consume too much design time. Chip Assemblers and Silicon Compilers are software tools which will significantly speed up "manual" chip construction. These tools are currently in a research state and will come to great importance in the future. Currently there are only three options to shorten design times of special purpose VLSI-chips: Gate Arrays, Standard Cells and General Cells.

All three approaches offer the extensive use of far reaching automatic placement and routing programs, promising very short chip design time.

As a system manufacturer, SIEMENS is working on all three approaches. Figure 2 shows the micrographs of a customized Gate Array (Gonauser, Mueller-Glaser, Glasl, 1979) used in high speed CPU's and a fully automatic designed standard cell chip for a Modem.

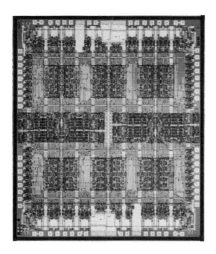 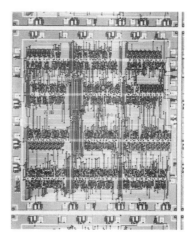

Fig. 2. Masterslice Gate Array with logic and RAM cells and a Standard Cell Chip for a Modem

Gate Arrays offer a maximum of 3000 to 5000 gate functions, they are limited in routing channels and they have no or only a very limited amount of ROM, RAM or PLA functions. The Standard Cell approach, characterized by cells with fixed height and variable width, is much more flexible in complexity of gate functions, chip size, placement and routing of cells, but the realization of on chip ROM, RAM or PLA within the given heigth of standard cells is not practible. Therefore both approaches are not suitable for real VLSI- chips. Only the General Cell approach, which allows placement and routing of cells of rectangular shape with variable heigth and width is suitable to VLSI, therefore the general cell approach has been chosen for the development of a comprehensive cell library.

Cell Library

To date the cell library contains 56 cells, most of them are equivalents to standard TTL-MSI functions, for example gates, multiplexers, barrel shifters, parity generators, comparators, flipflops, latches, registers and counters. Also some LSI cell functions like an 8-bit RALU Slice, an 8-bit Sequencer Slice (with AMD 2901/2909 functions as a subset), an adaptable size PLA and a 256 x 9 bit static RAM are included, as well as various MOS/TTL drivers. These functions were chosen for an easy transformation of pc-board logic into VLSI-chips.

The TTL equivalent cells are standard cells having uniform height and variable width with signal pads placed on one side. For these low complex functions the standard cell format is sufficient and makes placement and routing easier, in particular power distribution. The LSI equivalent functions on the other hand are general cells having a rectangular shape with power supply lines and signal pads distributed (placed) on all four sides. Figure 3 shows the 8-bit sequencer slice as an example for a general cell and the functional equivalent of the TTL 8-bit register 74377 as an example for a standard cell .

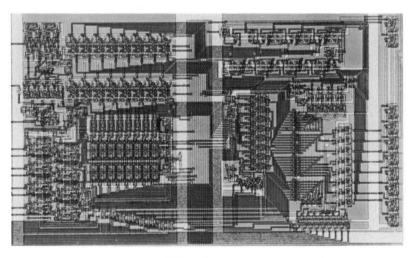

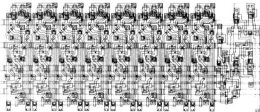

Fig. 3. 8-bit Sequencer Slice as a general cell. 8-bit Register (74377) as a standard cell.

All cells are designed with uniform geometrical and electrical input/output characteristics. In particular each cell output is equipped with a standard output driver to drive high capacitive loads. This type of standardization facilitates the use of automatic placement and routing to a considerable extent. All cells have gone through the production line on special testchips and have been tested for qualification before beeing released for production. Figure 4 shows one of the testchips.

Table 1 gives some statistics of the standard cells. The high number of transistors per gate function results because one gate function was counted independant of the number of inputs and the output driver (e.g. a 13-input NAND cell equals 1 gate function). Regularity within the standard cells, which is the ratio of transistors needed to transistors individually drawn, is 1.2 , so it is not astonishing that the designer's productivity for these optimized standard cells with all their constraints was only one third of that reported elsewhere for the design of a custom LSI chip (Mudge 1980). For LSI equivalent cells regularity is at least 8.

Table 1

Statistics of Standard Cells

	Min.	Max.	Average
gates/cell	1	60	12.6
transistors/cell	7	211	52.5
transistors/gate	2.22	32	4.17
gates/mm**2	37	588	268
transistors/mm**2	750	1908	1392
delay/gate	1.3 ns		
power/gate	500 uW		
delay/cell	5.2 ns with 1 pF load		

Technology

For this cell approach an advanced Silicon-Gate, N-Channel HMOS process developed at SIEMENS Integrated Circuit Division for the production of dynamic memories and highly sophisticated logic circuits has been chosen. The silicon gate process is defined by seven mask steps, a gate length of 3um and a gate oxide thickness of 50 nm. Enhancement device threshold and punch through are controlled by a double Boron implant in the channel region, while the depletion loads are provided by an Arsenic implant in the desired area. After the polysilicon is patterned, the source and drain regions are doped with Arsenic implantation providing self alignment with low underdiffusion and minimum overlap capacitance. Interconnection wiring is performed with diffusion, polysilicon and one layer of metal; the minimum linewidth is 2.5 um and minimum spacing is 3 um in the case of the shrink version, resulting in a high packaging density and in low power consumption. Figure 4 shows a testchip in normal (3.5um) and shrink version (3um), fabricated with a block reticle.

Fig. 4. Micrograph of a cell testchip with 32 standard cells and a 3x(256x9) bit static RAM.

CAD - System for custom VLSI

Figure 5 shows an overview of the PRIMUS CAD System used for designing VLSI chips (Zintl, 1981). Starting from the logic description in the common data base, a logic and timing simulator is used for verification of the logic design. Test patterns for this simulation can be written by hand or can be generated on a firmware level with the aid of a meta-assembler.

These patterns can also be converted by a Test Pattern Generator to functional patterns for wafer or chip tests with a SENTRY VIII Testsystem. For some applications test pattern generation is possible with a LASAR System. For finding static or dynamic failures inside the chip, an Electron Beam Testsystem is avaiable (Wolfgang et al., 1980).

Starting from the cell library described above, with allready functionally and electrically charcterized cells, VLSI chip layout is done interactively and to some extend automatically with the CALCOS System (Lauther, 1981). This system includes a program for placement and routing of standard cells, a program for placement and routing of general cells and a control program which checks the intercell wiring layout against the logic diagram. This control program also calculates electrical wiring data so that a final simulation including wiring delays is possible.

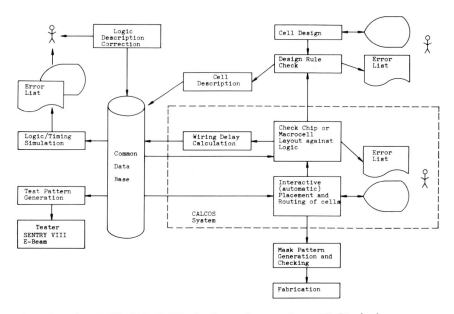

Fig. 5. The PRIMUS CAD System for custom VLSI design.

The CALCOS System allows the description of a placed and routed group of standard cells as a general cell, or a group of general cells as a macro cell, so that a recursive and hierarchical chip layout and design rule check is possible.

CAM programs for generating and testing masks for wafer fabrication have recently been added to this CAD System.

Applications

Currently the cell library and the whole CAD System is being tested on a pilot project, which is a special purpose data transmission processor called MIDUP. Table 2 gives some statistics of the MIDUP chip.

The first results, starting from the complete cell library and CAD System, indicates an estimated increase in productivity of a factor 5 - 10 on the base of an interactive layout compared to the traditional approach of fully optimized custom chip design.

However table 2 shows that the design time saved is partly paid back by an increase in the chip area, especially wiring area, wiring delays and power consumption. A further increase in productivity is possible with more automatic layouts. An attempt was made for a fully automatic placement and routing of the chip. It tooks 1200 cpu sec. on a SIEMENS 7.760 computer to place all cells and route more than 80% of the intercell wiring. But the result was not usable, because wiring delays on polysilicon interconnects were too high.

Table 2

Statistics of MIDUP chip

transistors overall	68300
logic only	23300
in standard cells	11900
number of cells	407
intercell signal lines	1600
chip area overall	79 mm**2
cell area	32 mm**2
wiring area	31 mm**2
power supply area	8 mm**2
power consumption	3500 mW
number of pins	132
regularity overall	34.15
logic only	12.15

3. CONCLUSION

Starting from a well designed library of general cells and standard cells with fairly high logic complexities and an integrated suite of appropriate CAD programs is a good basis for the economic design of customized VLSI circuits even with low production volumes. The results so far must be further improved by enhancing automatic layout. This is currently impeded by the high resistance of the polysilicon interconnects, which are not usable for power supply lines and give high delays for signals. Therefore a second metal interconnect layer is a must for this design philosophy.

ACKNOWLEDGEMENT

The authors wish to thank:
E. Baltin, T. Canzler, K. Ditschke, O. Doertok, D. Essl,
G. Geiger, K. Horninger, W. Joerger, K. Kling, E. Koesler,
U. Lauther, E. Schiller, H. Schulte, R. Sehr, J. Tacke
and W. Wach.

REFERENCES

Gonauser, E., Mueller-Glaser, K.D., Glasl, A. (1979). "Microprocessing Applications of Subnanosecond Masterslice Arrays". Siemens Forsch.- u. Entwickl.-Ber., Vol.8, No. 5.

Lauther, U. (1981). "The CALCOS System for Cell Based VLSI Design". to be published in "Hardware and Software Concepts in VLSI" (Ed. G. Rabbat). van Norstrad Reinhold Company.

Mudge, J.C. (1980). "Towards Structured Chip Design". NATO advanced summer institute on design methodologies for VLSI circuits.

Wolfgang, E., Fazekas, P., Otto, J., Crichton, G. (1980). Internal Testing of Microprocessor Chips using Electron Beam Techniques". Proceedings of the 1980 International Conference on Circuits and Computers".

Zintl, G. (1981). "A CODASYL CAD Data Base System" to be published in "Proceedings of the eighteenth Design Automation Conference 1981".

A SWITCH-LEVEL MODEL OF MOS LOGIC CIRCUITS

Randal E. Bryant

*Department of Computer Science,
California Institute of Technology, Pasadena,
CA 91125, USA*

INTRODUCTION

The study of mathematical models of logic circuits in recent times has been limited primarily to the Boolean logic gate model, in which a system consists of a set of unidirectional logic elements (gates) connected by one-way, memoryless wires. In contrast to this restricted model, designers of MOS LSI systems have a rich variety of circuit design techniques at their disposal. Combinational logic can be implemented with logic gates, steering logic and PLA's. Data can be stored in static and dynamic memory, communicated along wires and busses, and directed through pass transistors. Each of these techniques has numerous variations, and hence the designer can tailor a system design according to speed, density and architectural needs. The Boolean gate model lacks this richness, because it fails to reflect the basic structure of MOS systems in which the logic elements (i.e. the field-effect transistors) are bidirectional, and the wires (including the attached transistor gates) have sufficient capacitance to store information. As a consequence, computerized tools and analytic techniques based on the Boolean gate model provide limited assistance for the MOS designer. Many programs such as logic simulators extend the Boolean gate model with special logic elements and additional logic states, but these programs lack generality and accuracy as well as any formal mathematical basis.

In this paper a new logic model will be presented which more closely matches MOS circuit technology and hence can describe the logical behavior of a wide variety of MOS logic circuits in a very direct way. In this *switch-level* model a network consists of a set of nodes connected by transistor "switches", where each node has a state 0, 1 or X (for unknown or undefined), and each transistor has a state open, closed, or unknown. Transistors have no assigned direction of information flow and can be assigned different strengths to model their behavior in ratioed circuits. Nodes retain their states indefinitely in the absence of applied inputs, giving an idealized model of dynamic memory. Nodes can be assigned different sizes to model the effects of charge sharing in ratioless circuits. The switch-level model differs greatly from both Boolean logic gate and relay models in the way logic states are formed. In keeping with the concept of a *logic* model, however, both the transistor strengths and the node sizes may take on only discrete values, and the electrical operation of a circuit is modeled in a highly idealized way.

This model provides a formal bases for switch-level simulation programs such as the author's MOSSIM (Bryant, 1980, 1981b). These simulators have demonstrated the advantage of a switch-level logic model. They can accurately simulate a wide variety of MOS designs at speeds approaching those of logic gate simulators. Furthermore, since the simulation network corresponds closely to the electrical network, it can be derived from a specification of the mask patterns by a relatively straightforward computer program such as the one described by Baker and Terman (1980). The development of a mathematical model of switch-level networks has led to a simulation alogrithm which improves on the previous ones in its generality, accuracy and simplicity. This paper describes material presented in greater detail and with more rigor in Bryant (1981b).

NETWORK MODEL

A switch-level network consists of a set of nodes connected by transistor switches. The nodes are of two types: *input* nodes, labeled $i_1,\ldots,i_m$, and *normal* nodes, labeled $n_1,\ldots,n_n$. Input nodes provide strong signals to the system and are not affected by the actions of the network, much like voltage sources in electrical networks. Examples include the power and ground nodes Vdd and Gnd, as well as all connections to the chip through input pads. Normal nodes have states determined by the operation of the network, and these states are stored dynamically much as the storage of charge in capacitors. Each normal node is assigned a *size* from the set $K = \{\kappa_1,\ldots,\kappa_q\}$ to indicate its approximate capacitance relative to other nodes with which it may share charge. The elements of K are totally ordered from κ_1 to κ_q. These node sizes allow a simplified model of charge sharing in ratioless circuits in which the states of the largest node(s) dominate when a set of nodes is connected by turned-on transistors. Figure 1 shows a switch-level model of a three transistor dynamic RAM circuit in which the bus node has size κ_2 to indicate that it can supply its state to the storage node of the selected bit position during a write operation and to the drain node of the storage transistor during a read operation. Most MOS designs can be modeled with just two different node sizes ($q=2$).

Each normal node n_j has a state $y_j \in \{0,1,X\}$. The states 0 and 1 correspond to the normal Boolean logic states, while the state X indicates that the node has not been properly initialized or that its voltage may lie between the logic thresholds due to either a short circuit or improper charge sharing. Each input node i_j has a state x_j with the same interpretation. The state of a network is given by two vectors x and y indicating the states of the input nodes and normal nodes, respectively.

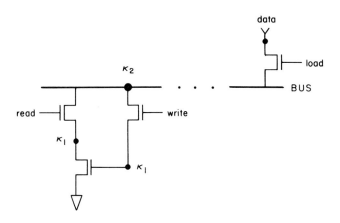

Fig. 1. Switch-Level Model of 3-Transistor Dynamic RAM

A transistor is a three terminal device with terminals labeled "gate", "source" and "drain". It acts as a switch with state determined by the transistor type and the state of the gate node as shown in the following table. The d-type (for "depletion") transistor is used to model both pullup load transistors in depletion mode nMOS circuits and the polysilicon-diffusion layer crossovers seen in some designs.

n-type		p-type		d-type	
gate	effect	gate	effect	gate	effect
0	open	0	closed	0	closed
1	closed	1	open	1	closed
X	unknown	X	unknown	X	closed

Each transistor is assigned a *strength* from the set $\Gamma = \{\gamma_1,\ldots,\gamma_p\}$ to indicate its approximate conductance when turned-on relative to other transistors which may form part of a ratioed path. The elements of Γ are totally ordered from γ_1 to γ_p. These transistor strengths allow a simplified model of ratioed circuits in which a path to an input node containing only conducting transistors of strength greater than or equal to some value overrides any path containing a transistor of strength less than this value. Figure 2 shows a switch-level model of an nMOS Nand gate with a pass transistor on its output. Most MOS designs can be modeled with just two transistor strengths ($p=2$), with the load transistors having strength γ_1 and all other transistors having strength γ_2, although some circuits involve multiple levels of ratioing and hence require more transistor strengths ($p \geq 3$).

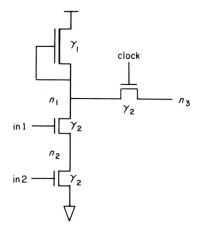

Fig. 2. *Switch-level model of nMOS Nand gate.*

THE TARGET STATE FUNCTION

The logical behavior of a switch-level network is characterized by its *target state function*. For a given set of input node states x and normal node states y, the target state y' is defined as the node states which the normal nodes would eventually reach if all transistors were held fixed in states determined by the initial node states. This function describes how the normal nodes attain new logic states due to connections to input nodes or other normal nodes through paths of conducting transistors. This definition ignores the fact that the transistors will change state in response to the changing states of their gate nodes. Thus the target state function only gives an indication of the instantaneous behavior of the network.

The target state function of a switch-level network closely resembles the *excitation* function of a logic gate or relay network, which is defined to give the set of states which would form at the outputs of the logic elements (logic gates of relay coils) in response to the network state given as argument. Huffman (1954) first recognized the importance of the excitation function for characterizing the logical behavior of a network, although he expressed it in terms of a *flow table* containing the excitation state for each possible network state. Thus, although the switch-level model differs greatly from both relay and logic gate models in the way logic states are formed, these models describe the logical behavior of systems in similar ways.

Given a means of computing the target state function, one can implement a form of "unit-delay" logic simulator which simulates the operation of a network by repeatedly applying the target state function. That is, with input nodes set to some state x and normal nodes set initially to state y, the network is simulated until it stabilizes in a state

$$y'' = \lim_{k \to \infty} T_x^k(y)$$

where the function T_x denotes the target state function for input state x, and the superscript k denotes k applications of the function. For most networks of interest, a stable state will be reached after a bounded number of iterations. Such a method is used by MOSSIM to simulate the effect of each change in clock or data input states. This simulation technique presents the user with a timing model in which the transistors switch one time unit (i.e. one application of T_x) after their gate nodes change state. Such a timing model has proved adequate for testing many LSI designs. Thus a method for computing the target state function provides the key to applying the switch-level model.

LOGIC SIGNALS

For a network containing no transistors in the unknown state,

the formation of logic states on the nodes can be described
in terms of an abstraction we call *logic signals*. A logic
signal has both state and strength, describing the dominating
effect of a network (or subnetwork) at some node, and the
relative capacitance or conductance of this effect.

Three types of signals describe the different effects a
subnetwork may have on a node. A *charging* signal has state
in the set {0,1,X} and strength in the set K. It indicates
a connection to a set of normal nodes with maximum size
equal to the signal strength. A *driving* signal has state
in the set {0,1,X} and strength in the set Γ. It indicates
a set of paths through conducting transistors to input nodes
with path strengths equal to the signal strength, where the
strength of a path equals the minimum transistor strength in
the path. Charging and driving signals of strength s and
states 0,1 and X are denoted -s, +s and xs, respectively.
Finally, a *null* signal, denoted λ, has a null state N and
strength 0. It indicates an open circuit. The set of signal
strength values is totally ordered

$$0 < \kappa_1 < \ldots < \kappa_q < \gamma_1 < \ldots < \gamma_p$$

indicating that a path to an input node can override a path
to a normal node, while either of these can override an open
circuit.

Using the set of signal values as domain, we can develop
an algebra describing the effects of performing some
elementary network transformations. First, when subnetworks
described by signals a and b are connected together at a
node, the net effect is described by a signal $a \lor b$ equal
to the least upper bound of a and b for the partial ordering
shown in Fig. 3. That is, a stronger signal will override a
weaker, while signals of the same strength will form a signal
of this strength and with the same state if they are equal
and state X if they conflict. Observe that the set of signal
values along with this partial ordering forms a lattice.

Second, when a subnetwork described by a signal a is
connected to a node through a transistor in the closed state
and having strength s, the net effect is described by a
signal $s \circ a$ with state equal to the state of a and strength
equal to the minimum of s and the strength of a. That is,
a charging signal will connect through unchanged, while the
strength of a driving signal may be decreased by the
connection. We will adopt a convention that $0 \circ a = \lambda$ i.e. a
connection through a 0 conductance gives an open circuit.

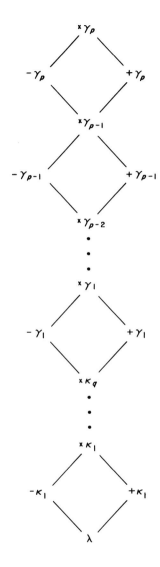

Fig. 3. The Lattice of Logic Signals

EQUATIONS FOR THE TARGET STATE

We wish to develop an equation specifying the target state of a network for a particular input node state x and initial node state y. For the special case in which the network contains no transistors in the unknown state, the effect of the network on the normal nodes is described by their *steady state signals*, denoted with the vector v, each having state equal to the target state of the node. The values of these signals can be expressed with a matrix equation as follows.

The signal formed by input node i_j, denoted x_j, has state x_j and strength γ_p, the maximum possible strength. The signal formed by normal node n_j, denoted y_j, has state y_j and strength equal to the size of the node. These sets of signals are denoted by the vectors x and y, respectively. The transistors in the closed state are described by two matrices G and E, with g_{ij} equal to the maximum strength transistor in the closed state connecting normal nodes n_i and n_j (or 0 if no such transistor exists), and with e_{ij} describing the analogous connection between normal node n_i and input node i_j. The set of steady state signals v must be the set of minimum values satisfying the equation

$$v = E * x \vee y \vee G * v \qquad (1)$$

In this equation $*$ denotes a matrix product with $\circ$ as the analog of multiplication and $\vee$ as the analog of addition. Furthermore, any time a scalar function such as $\vee$ is shown applied to vector arguments, we mean its pointwise extension to a function which yields a vector with elements equal to the result of applying the scalar function to the corresponding elements of the argument(s). This equation expresses the fact that the effect of the network on node n_i equals the combined effect (i.e. least upper bound) of the initial charge on the node as described by the signal y_i; the direct connections to each input node i_j as described by the signal $e_{ij} \circ x_j$; and the connections to the rest of the network through each other normal node n_j as described by the signal $g_{ij} \circ v_j$. Moreover the set of steady signals equals the set of minimum values satisfying all of these constraints. Observe that this equation has the form $v = f(v)$ with v equal to the least fixed point of the function f.

Equation 1 only applies to networks containing no transistors in the unknown state. In general, however, a network may contain n-type or p-type transistors with gate nodes in the X state. Since this state represents an unknown node voltage anywhere between 0.0 and V_{dd}, the transistor will have an unknown conductance anywhere between 0 and the conductance when fully turned-on. The target state of a node is defined to equal 0 or 1 if and only if it will have this unique state regardless of the conductances formed by the transistors in the unknown state, and otherwise the target state equals X. This definition seems to require trying a possibly exponential number of cases with the transistors in the unknown state set to all possible combinations of open and closed. Fortunately, the target state of an arbitrary switch-level network can be expressed by a set of matrix equations in an algebra of signal strengths as follows

$$\mathbf{r} = \mathbf{E}^{\min} \cdot \|x\| \uparrow \|y\| \uparrow \mathbf{G}^{\min} \cdot \mathbf{r} \qquad (2a)$$
$$\mathbf{u} = block(\mathbf{E}^{\max} \cdot \lceil x \rceil \uparrow \lceil y \rceil \uparrow \mathbf{G}^{\max} \cdot \mathbf{u},\mathbf{r}) \qquad (2b)$$
$$\mathbf{d} = block(\mathbf{E}^{\max} \cdot \lfloor x \rfloor \uparrow \lfloor y \rfloor \uparrow \mathbf{G}^{\max} \cdot \mathbf{d},\mathbf{r}) \qquad (2c)$$

In these equations, for a signal a, $\|a\|$ equals the strength of a, $\lceil a \rceil$ equals the strength of a if a has state 1 or X and equals 0 otherwise; and $\lfloor a \rfloor$ equals the strength of a if a has state 0 or X and equals 0 otherwise. The operation $\uparrow$ gives the maximum of its arguments and $\cdot$ denotes a matrix product with the minimum function as the analog of multiplication and $\uparrow$ as the analog of addition. For two strength values a and b, $block$ (a,b) equals a if a $\geq$ b and equals 0 if a < b. The matrices $\mathbf{G}^{\min}$ and $\mathbf{E}^{\min}$ describe the minimum possible connections between the nodes in which transistors in the unknown have 0 conductance. The matrices $\mathbf{G}^{\max}$ and $\mathbf{E}^{\max}$ describe the maximum possible connections between the nodes in which transistors in the unknown state are fully conducting.

For a vector **r** equal to the minimum solution of equation 2a, each element r_i equals the strength of the steady state signal for node n_i when all transistors in the unknown state have 0 conductance. For vectors **u** and **d** equal to the minimum solutions of equations 2b and 2c, each element u_i equals the strength of the strongest possible steady state signal on node n_i having state 1 or X for any combination of conductances formed by transistors in the unknown state, while each element d_i equals the corresponding value for signals with states 0 or X. A node n_i will have target state 1 if and only if no possible combination of transistor conductances could give a signal on n_i of state 0 or X, which implies that $d_i = 0$, and similarly it will have target state 0 if and only if $u_i = 0$. Thus the target state can be computed as

$$y'_i = \begin{cases} 1, & d_i = 0 \\ 0, & u_i = 0 \\ X, & \text{else.} \end{cases}$$

Thus we have a specification of the target state for an arbitrary network.

COMPUTATION OF THE TARGET STATE

Equation 2a has the form of a fixed point equation $\mathbf{r} = f_s(\mathbf{r})$, where the function f_s is monotonic. The set of signal strengths is finite and totally ordered, and hence it forms a continuous lattice, and any monotonic function over it is continuous, as defined by Scott (1972). Thus, Scott's theorem regarding the least fixed point of a continuous function on a continuous lattice can be applied to show that equation 2a can be solved by an iterative technique where $\mathbf{0}$ denotes a vector of all 0's.

$$\mathbf{r} = \lim_{k \to \infty} f_s^K(\mathbf{0})$$

Furthermore, convergence will be reached in a bounded number of steps. Using this vector in equation 2b and 2c then gives fixed point equations for $\mathbf{u}$ and $\mathbf{d}$ which can be solved by the same method.

As an example of the computaion of the target state, suppose that the network of Fig. 2 has inputs in1 = in2 = 1 and clock = X, and that node n_3 has size κ_1 and initial state 0. The recurrence equation for $\mathbf{r}$ can be written as

$$r_1 = \gamma_1 \uparrow (\gamma_2 \downarrow r_2)$$
$$r_2 = \gamma_2 \uparrow (\gamma_2 \downarrow r_1)$$
$$r_3 = \kappa_1$$

where the operation $\downarrow$ gives the minimum of its arguments. Applying the iterative method the following sequences of values:

$$r_1: \quad 0 \quad \gamma_1 \quad \gamma_2 \quad \gamma_2 \quad \cdots$$
$$r_2: \quad 0 \quad \gamma_2 \quad \gamma_2 \quad \gamma_2 \quad \cdots$$
$$r_3: \quad 0 \quad \kappa_1 \quad \kappa_1 \quad \kappa_1 \quad \cdots$$

from which we get a solution $r_1 = r_2 = \gamma_2$, and $r_3 = \kappa_1$. The recurrence equation for $\mathbf{u}$ can be written as

$$u_1 = block(\gamma_1 \uparrow (\gamma_2 \downarrow u_2) \uparrow (\gamma_2 \downarrow u_3), \gamma_2)$$
$$u_2 = block(\gamma_2 \downarrow u_1, \gamma_2)$$
$$u_3 = block(\gamma_2 \downarrow u_2, \kappa_1)$$

which has a minimum solution $u_1 = u_2 = u_3 = 0$, indicating that regardless of the conductance formed by the pass transistor, no signal with state 1 or X can form on any nodes. Even though the pullup transistor provides a signal of strength $+\gamma_1$ to node n_1, our computation correctly recognizes that this signal will be overridden by the signal $-\gamma_2$. The recurrence equation for $\mathbf{d}$ is

$$d_1 = block((\gamma_2 \downarrow d_2) \uparrow (\gamma_2 \downarrow d_3), \gamma_2)$$
$$d_2 = block(\gamma_2 \uparrow (\gamma_2 \downarrow d_1), \gamma_2)$$
$$d_3 = block(\kappa_1 \uparrow (\gamma_2 \downarrow d_2), \kappa_1)$$

which has a minimum solution $d_1 = d_2 = d_3 = \gamma_2$. Thus, since these values are all nonzero, while the values of $\mathbf{u}$ are all 0, all three nodes have a target state 0.

If the same network has initial state 1 for n_3, we would find that $u_3 = \kappa_1$, while all other elements of $\mathbf{u}$ and $\mathbf{d}$ have the same values as before. This gives target states $y'_1 = y'_2 = 0$, and $y'_3 = X$, indicating that the unknown conductance of the pass transistor creates an ambiguity in the target state of node n_3.

A logic simulator can be implemented which repeatedly computes the target state using this method. By exploiting the spareseness of the network and the fact that the activity in a network is highly localized, this simulator can operate at speeds comparable to traditional logic gate simulators.

CONCLUSION

The formal model of switch-level networks provides a mathematical link from the physical structure of an MOS system to its logical behavior. This has direct applications in the area of logic simulation, giving logic simulators with greater expressive power and accuracy than those based on the Boolean gate model. Furthermore, other computerized tools and analytic methods can benefit from this ability to move between these two different views of a system.

ACKNOWLEDGEMENTS

This research was supported in part by the United States Department of Energy under contract number DE-AC02-79ER10473, and in part by United States Air Force Contract AFOSR F49620-80-C-0073.

REFERENCES

Baker, C.M. and Terman, C. (1980). Tools for verifying integrated circuit designs, *Lambda Magazine*, 4th Quarter, pp. 22-30

Bryant, R.E. (1980). An algorithm for MOS logic simulation, *Lambda Magazine*, 4th Quarter, pp. 46-53.

Bryant, R.E. (1981a). "A Switch-Level Simulation Model for Integrated Logic Circuits". Phd Thesis, MIT Dept. of EECS.

Bryant, R.E. (1981b). MOSSIM: a switch-level simulator for MOS LSI, *Proceedings, 18th Design Automation Conference*.

Huffman, D.A. (1954). The synthesis of sequential switching circuits, *Journal of the Franklin Institute* **257**, pp. 161-190, 275-303

Scott, D.S. (1972). Continuous lattices. *In* "Toposes, Algebraic Logic and Logic" (Ed. F.W. Lawvere). Springer-Verlag, Berlin

FAILURE MECHANISMS, FAULT HYPOTHESES AND ANALYTICAL TESTING OF LSI-NMOS (HMOS) CIRCUITS

B. Courtois

*Laboratoire Informatique et Mathematiques Appliquees
b.p. 53 X, 38041 Grenoble Cedex, France*

ABSTRACT

This paper is concerned with a classification of failure mechanisms, followed by its application to both electrical and logical levels for N-MOS (H-MOS) technology. A link between failure analysis, from reliability physics, and test design is thus carried out. Classes for fault hypotheses are given as a function of possible occurrences of failure mechanisms, and of the ease of test generation. Test vector generation methods are illustrated, both at gate level, and at the level of typical complex circuits such as ALU, ROM and PLA. The classes of hypotheses are well defined. The future developments of analytical testing are discussed together with the evolution of the design automation of VLSI parts.

KEY-WORDS and PHRASES

Test, analytical testing, fault hypotheses, failure mechanisms, VLSI design.

1. INTRODUCTION

The stuck-at model has been used for several years as a fault hypothesis in logical circuits. Friedman and Menon have noted [FRI 71] that most failures in DR, DTL, TRL, and TTL technologies are failures that may be modelled as stuck-at failures of an input or an output. Chang, Manning and Metze in [CHA 70] have given examples of stuck-at faults, but note that problems in testing are mostly due to the techniques for test generation rather than the accuracy of the mode in reflecting real failures.

Other types of failures mentioned in the literature are the shorted input diodes, input bridging faults, feedback bridging faults, delays. An interesting study has been made by Crouzet, Galiay, Rousseau and Vergniault in [CRO 78]. A comparison is made between the stuck-at model and opens and shorts in diffusion and metallization for gates made from a load device and a net of transistor switches. Some possible failures cannot be modelled by a stuck-at model, and vice versa.

The aim of the present study is multiple. Failure mechanisms have been investigated because their effects are not always understandable by logicians. The second step is to relate them to the electrical and logical levels. Another aim is to take into account potential failures that may appear during the life of the circuits (not all failures that may result from the manufacturing process).

Examples of gates not having an equivalent logical form (load and switch net) will be given. Some general results obtained for regular circuits will also be given.

2. FAILURE MECHANISMS OF INTEGRATED CIRCUITS

It is not easy to give a classification of failure mechanisms, because of the mixing of causes and effects. Several causes may produce the same effect. A single cause may give several different effects depending on the presence of other parameters. The classification we have chosen corresponds to the work "International Classification for Physics" [ICSU/AB]. The last section proposes different sections of classification.

2.1 *Transport phenomena in solids*

Transport phenomena result generally from the diffusion of particles in a medium. The medium is made up of metals and alloys. Outwith an electrical field, the important

parameters are the concentration gradients and the temperature. An electrical field introduces other factors:
. an electrostatic force acting on positive and negative ions;
. a force due to the electronic wind.

This is the phenomenon of electromigration. Phenomena affecting the metallization and the connections between die and package have been separated.

In metal, several entities may migrate : atoms of the metal, impurities and faults. Typical consequences are hillocks, whiskers and pits, and voids. In [COU 80] some examples are given for systems like AL- Si. Step coverage and misalignment encourage these failures. Interdiffusion has been largely discussed in the past.

2.2 Surface growth : dendrites and whiskers

Dendrites and whiskers growth has been generally attributed to electrochemical problems. These growths are due to instability problems in interfaces. They may appear without electrochemical action. In this case, material is coming from the inside of the crystal itself. These phenomena are for example responsible for reconnection problems in PROMs.

2.3 Dielectric and conductor breakdown

Breakdown of SiO_2 films is primarily due to the emission of electrons by the cathode. This leads to local heating, which activates positive ions. These move towards the cathode and increase the field, etc. The positive ions are Na^+ and H^+ or $H3O^+$ from NaCl and H_2O. Metal fuses in cases of overstress.

2.4 Thermomechanical effects

Thermomechanical effects are due to expansion coefficients, that are different, for example, for silicon and ceramic. The shape of wires is a parameter for these effects.

2.5 External agents

In this section failure mechanisms have been grouped with respect to the action of external agents. These agents may be present initially in the package (but not active), or may penetrate in the package. The consequences of ionic contamination are easy to understand. The characteristics of gates are modified. Oxide defaults like pinholes, stacking faults, etc. increase these effects.

Another external agent is energy which may be due to radiation, this is widely studied for RAMs, or to vibration, e.g. ultrasonic vibrations.

Chemical corrosion is a major contaminant of circuits. The constituents are humidity, soluble salts and an electrical field. Beside the electrochemical consequences, the process begins by the dissociation of H_2O and continues by the constitution of substances that attack metals.

In that brief survey, some mechanisms have not been listed such as degree errors, e.g. a parasitic MOS, or mechanisms specific to manufacturing process, diffusion shorts by pipes and strikes, shorts between poly Si and Al. Finally the thermal noise is not taken into account because of the large dimensions of circuits.

3. FAILURE MODES AND FAULT HYPOTHESES FOR N-MOS CIRCUITS

3.1 *Electrical failure modes*

Table 1 summarizes electrical failure modes due to the above listed failure mechanisms. Failures affecting the die and the package are distinguished in silicon gate NMOS technology.

3.2 *Fault hypotheses (classes)*

Three classes of fault hypotheses have been considered. These classes reflect an increasing "quality" of testing, in addition to logical hypotheses made in the past.
Multiple stuck-at are a refinement of single stuck-at faults
. Class 0
This class is the class of the *single physical* defects, such as one failed contact, one failed MOS, one cut. Shorts are not included in this class : two defects are supposed to be necessary to create a short. A floating gate is supposed to result in a stuck transistor. The single defect notion is of course relative to a reference circuit.
. Class 1
This class is an extension of class 0. Some shorts are to be tested : shorts between an Al metallization and the geographically nearest metallization. Similarly, because of the same complexity for test, shorts between diffusions are tested.

Table 1 - Electrical failure modes for N-MOS technology

	Localisation	Physical failure mechanism that may be evoked	Electrical failure mode
Cuts of inter-level contacts	Contact Al - poly Si Contact Al - diffusion Precontact poly Si - diffusion	Transport phenomena (mechanical weakness)	Open circuit Open circuit (Open circuit)
Cuts of interconnections lines	Al Poly Si Diffusion	Breakdown (mechanical weakness) Ionic contamination	Open circuit (Open circuit, floatting gate) Open circuit
Shorts between elements of a level	Al Poly Si Diffusion	Electrochemical phenomena Manufacturing only (Ionic contamination)	Shorts (Shorts)
Shorts between near levels (vertically)	Al - Poly Si Si - diffusion	Manufacturing only Overstress breakdown (girth) Ionic contamination (die)	MOS s-on, s-open
Shorts between a vertical element and another near level	Contact Al - Diffusion and near Poly Si	Manufacturing only	
Cuts or shorts of connections die-package	Bonds Wires Leads	Inter-diffusion Thermal weakness Corrosion	Open circuit Open circuit Open circuit
Shorts between connections die - package	Wires, leads	Particles	Short

Class 2

Class 2 includes what has been excluded from class 1. Shorts between metallization are considered. Multiple shorts are tested. Shorts between diffusions are similarly considered. Multiple defects are tested. Only the failures supposed to be due to the manufacturing process, e.g. shorts between metallization and diffusion are excluded.

The three classes are summarized in Table 2. Class 0 represents a minimal test. Class 1 represents a more difficult test. Class 2 is a "deluxe" test.

Table 2 - Fault hypotheses

Class 0	Class 1	Class 2
One physical single defect : 1 contact, 1 precontact, 1 MOS s-on or s-open, 1 Al cut ..... floating gate ==> MOS s-	Class 0, plus : 1 short between 1 Al and the (physically) nearest Al. Idem for diffusions.	Class 1, plus : shorts between any Als. Idem for diffusions. multiple defects.

4. LOGICAL FAILURE MODES AND ANALYTICAL TESTING

The aim of this section is to illustrate the necessity of the fault hypotheses derived above and the way to derive tests according to the fault hypotheses. The way to handle regular arrays will be presented.

4.1 Classical and complex gates

It is very easy to see that class 0 and the logical stuck-at model are equivalent for an elementary NOR gate. A classical gate, composed of a load transistor and a set of switches, shows the impossibility of considering a logical stuck-at model if the function is not of the standard form, $\Sigma \Pi$. This is illustrated on Fig 1. Figure 2 represents a conjunction gate, used in the Z80, exhibiting a class 0 defect which cannot be modelled by the logical stuck-at model. For that gate, an examination of class 0 defects reveals that a failure affecting the central load transistor would only be detected by a 11 vector when test vectors derived from stuck-at hypotheses of the inputs or of the output would be (00, 01, 10) or (01, 10, 11). A similar example is represented on Fig 3. That gate is used in the 9511 circuit. Testing the inputs and outputs would be possible using the vectors (00, 01, 10). But only the vector 11 tests class 0 defects indicated on Fig 3.

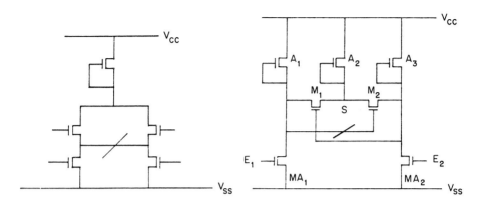

Figure 1 : Classical gate exhibiting a class 0 defect which cannot be modelled by a logical stuck-at model.

Figure 2 : Conjunction gate exhibiting a class 0 defect which cannot be modelled by a logical stuck-at model (Z 80).

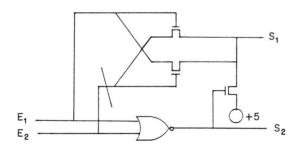

Figure 3 : Gate exhibiting a class 0 defect which cannot be modelled by a stuck-at model

4.2 Analytical testing

Of course, the main problem is how to derive "logical" test vectors according to the fault hypotheses of Table 2. Accordingly, an example is given and regular arrays are examined.

4.2.1 Electrical path sensitizing

Path sensitization is a commonly used method to derive test vectors from a logical stuck-at model. That method may be revised at the electrical level. Testing, for example, a class 0 defect may be possible using sensitized paths including the failed element or, using a path with a command path. A sensitized path includes one or several current sources and one or several grounds. It is sensitized when the tested element may be observed by a variation of one or more outputs, by means of blocking parallel paths and opening "serial" elements. This is illustrated below using the conjunction gate represented in fig 2. To test the transistor E_2 by means of two sensitized paths:
- the transistor M_2 being blocked on by $E_1 = 0$, the sensitized path is made up of the source A2, M1, controlled by E1, ground MA1. That path is controlled by E_2.
- The transistor M_2 being blocked on by $E_1 = 0$, the sensitized path is made up of the sources A2 and A3, of the transistor controlled E_2, and of the ground MA2.

The tests of the transistors M1 and M2 are similar to the tests of the transistors controlled by E_1 and E_2. The test of the source A2 requires the opening of M1 and M2 by $E_1 = E_2 = 1$. The test of the source A1 is made using the sensitized path A2, M2, E2.

The determination of test vectors for all defects of class 0 allows the determination of test vectors for the gate, in a similar way as for logical test hypotheses. To determine test vectors for class 1 fault hypotheses requires the physical design of the circuit to be taken into account.

In [COU 81.1] may be found the test derivation for the data processing section of the M6800, including the tests of operators, using iterative properties. Other circuits are interesting for analytical test derivation : the ROM and PLA arrays.

4.2.2 Regular arrays

Path sensitizing may be used to derive tests for ROMS and PLAs very easily. These circuits are widely used in control sections, e.g. in the Z80, 8085, 8748. The test of PLAs has been studied by many authors. The hypotheses generally made give two types of consequences:
. failures in the AND plane or in the input decoder may result in the abnormal activation of a minterm, or in the

abnormal non activation of a minterm, or in both abnormal situations
. failures in the OR plane may result in an increase or decrease in minterm combinations. The fault hypotheses of Table 2 restricts testing to failures that may effectively occur.

For the first and second planes, tests of class 0 and 1 have been developed and may be found in [COU 80]. For the complete PLA array, a class 0 test needs a number of tests less than the sum of the transistors implemented on the array, and of the number of outputs of the first and second planes. A class 1 test needs to add a number of tests less than the sum of the number of outputs of the two planes. These tests are not detailed here because of lack of space.

5. ANALYTICAL TESTING AND VLSI

Fault hypotheses as reported in Table 2 have to be refined in several ways. One way is to take care that some multiple defects are tested by class 0, in the same way that the test of single stuck-at faults will test some multiple stuck-at faults of a logical model. Another way is to extend the classes of Table 2 by considering shorts between not only nearest Al metallization, but also because one metallization and a second nearest one, etc....

Other process technologies need to be studied. For example, a two level poly Si would need to take into account other specific failure mechanisms.

Of course, just regular circuits may be analytically tested. For example, the control section of the M6800 has to be functionally tested. Fortunately, for on-line testing without massive redundancy [COU 81.2], just the analytical testing of operators is needed, and these operators exhibit regularity properties.

A last remark is directed to the development of analytical testing together with VLSI design evolution. It is clear that the evolution is towards structured design. In this framework, PLAs are one example. More generally, the evolution is towards high level cells: ROM, RAM, ALU slice [SUZ 81]. This allows the test of individual cells, and the assembly of tests for an entire circuit. Test derivation could be course be included in CAD systems, helping the design of these cells. Test derivation for PLAs could be included in logic minimization programs, which are necessary for the design of large systems. Specific types of circuits like systolic arrays exhibit a higher level of structure assisting the design of tests.

REFERENCES

[CHA 70] CHANG H.Y. *et al.*, "Fault diagnosis of digital systems", Wiley Interscience, 1970.
[COU 80] COURTOIS B., "Mecanismes de pannes, hypotheses de pannes et test analytique de circuits LSI N-MOS (H-MOS)" Internal Report IMAG RR N° 196, April 1980, 100 pp.
[COU 81.1] COURTOIS B., "Analytical testing of data processing sections of integrated CPUs", To Appear *in* International Test Conference, Cherry Hill, Philadelphia, October 27-29 1981, USA.
[COU 81.2] COURTOIS B., "A methodology for on line testing of microprocessors", 11th Fault Tolerant Computing Symposium, Portland, June 24-26 1981, USA.
[CRO 78] CROUZET Y. *et al.*, "Improvement of the testability of LSI circuits", Proceedings 4th European Solid State Circuits Conference, Amsterdam, The Nederlands, 1978.
[FRI 71] FRIEDMAN A.D. & MOMON P.R., "Fault detection in digital circuits", Prentice Hall Inc., 1971.
[ICSU/AB] Abstracting board of the International Council of Scientific Unions.
[SUZ 81] SUZIM A.A., "Data processing section for microprocessors-like integrated circuits", IEEE Transactions on Solid-state circuits, June 1981.

ACKNOWLEDGEMENTS:

The study reported herein has been possible because of the VLSI design studies of the Computer Architecture Group of the IMAG Laboratory. This work has been supported by CNET under grants nos 76B378 and 799B085 (CANOPUS Project).

Automatic Synthesis, Verification and Testing

by J. Paul Roth

*IBM T. J. Watson Research Center,
Yorktown Heights NY 10598 USA*

ABSTRACT. In VSLI there is acute need for automatic logic design, verification and testing. Described here is: 1) a high-level hardware language PL/R, an infinitesimal variant of PL/I which can be simulated and whose compiler RTRAN produces Logic largely comparable to manually produced Designs; 2) methods for verifying COMPLETELY the equivalence/inequivalence of two logic designs or of a design against a high-level specification; this method was used comprehensively in the design of an IBM Processor Complex; 3) an extension of the method to the verification of engineering changes by employment of the D-algorithm, is described which reduces running time by an order of magnitude (in each case a counterexample - usually many of them - is produced in the event that they are not equivalent); 4) new methods of logical design and of execution of the D-algorithm which avoids the use of LSSD, at the expense of extra computation (in large designs usually a great deal of computation); 5) by imposition of a condition of STRONG REGULARITY, partitioning the design via registers connected in shift-register mode, the methods of verification and testing can be extended from 100,000 to 1,000,000 circuits and more.

INTRODUCTION. VLSI presents new problems of complexity, for logic design, for testing and verification, to say nothing of the problem of physical design - the layout and wiring. Given here are methods of automatic logic design which, for control logic, is comparable to good manual design - comparable arithmetic design is more difficult. Given also are methods for verifying completely a logic design either against another detailed design or against a high-level specification thereof. Also a new method of design which obviates the need for LSSD (level sensitive scan design (1980) for diagnosis, at the expense of considerably more test-generation computation and simulation. Finally a method is given for partitioning a logic design which converts a VLSI diagnosis job into an LSI diagnosis job.

THE METHODS. PL/R is a language for high-level logic design (1980). It is isomorphic to an extremely small portion of PL/I, having the logical functions of And, Or and Not, operating on vectors, the DO operation and macro insertion. It was developed originally by Harry Halliwell of IBM UK at Hursley. A compiler RTRAN was developed from the PL/I optimizing compiler which transformed a PL/R program into a REGULAR logical design RLD, having for cyclic logic, registers in the LSSD format.

For example the PLR version of a three-input majority function would be written

=MAJ(A,B,C,M); INPUT A,B,C; OUTPUT M; M=A&B | B&C | C&A;

A 16-bit stripped-down arithmetic and logic unit ALU16 is next given in PL/R.

=ALU16(R,S,T,A,O,X,P,KI,KO);
INPUT R(16),S(16),A,O,X,P,KI;
OUTPUT T(16),KO;
DCL TA(16),TO(16),TX(16),TP(16),K(16);
TA=R&S;
TO=R | S;
TX=¬R&S | ¬S&R;
K(0)=KI;
DO J=0 TO 14;
K(J+1)=K(J)&TO(J) | ¬K(J)&TA(J);
END;
TP=¬TX&K | ¬K&TX;
T=TA&A | TO&O | TX&X | TP&P;
KO=K(15)&TO(15) | ¬K(15)&TA(15);

Here the primitive operations of vector Oring TO, Anding TA and Exclusive-oring TX are first defined. Then the Carry K is defined by means of a DO loop. These two PL/R designs produced in their RTRAN implementations respectively 13 circuits and 521 circuits. 2) VERIFICATION. There are two types of verification available for hardware. In the first case a new design is compared comprehensively with an old design presumably having the same functions. In the second a new design is verified against a high-level specification of itself. There must be a one-to-one correspondence between primary inputs and primary outputs of the two designs. In addition if we have a REGULAR design- into each feedback loop is inserted pair of registers, gated at sufficiently different clock times that races and hazards are avoided - we also require a one-to-one correspondence between the pseudo-inputs and -outputs of the registers (feedback loops). This greatly reduces the verification task.

Let us consider the case of the verification of a design against a full specification of the design; it will be seen that a comparison of two designs may be handled as a special case. Consider the diagram in Fig. 1

Here at the top is represented a functional specification of the design, presumably written in PL/R or any functional specification. It is assumed that the entire design is so defined. RTRAN or any other hardware compiler maps these specifications into an automatic design. At the same time designers, perhaps hundreds of them, will implement the design manually. Then the two designs, one manual, one automatic, are compared one against the other. VERIFY was used for this purpose in the design of the IBM processor complex 3081, together with RTRAN. This system for verification has the capacity of catching many, many design errors and inconsistencies. These mistakes are caught before fabrication. Without such a system VLSI computer design would be prohibitively expensive and time-consuming.

The output of VERIFY would be either a counterexample - usually many of them - or else a proof of equivalence. 3) VERIFICATION VIA THE D-ALGORITHM. Imagine in the process of a design an engineering change EC is made. There will be a set of outputs which are supposed not to change in function with respect to the EC; the EC is the common form of change in design - perhaps thousands of them. Question: how efficiently verify engineering changes?

The D-algorithm may be used if the change be treated as a FAILURE. If the change is expressed as a number of "Adds" and "Deletes", then the P*-algorithm may be used (1980) to compute the "primitive D-cube of failure" for the change. Then *A LA* the D-algorithm distinguishing tests, if they exist, can be computed. This method is at least an order of magnitude

Fig. 1 **HARDWARE VERIFICATION**

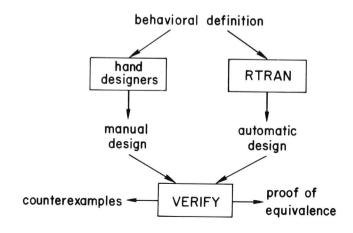

faster than methods such as VERIFY which assume no information concerning the similarity of the design.

5) VLSI Diagnosis. Imagine a design with a million or more circuits. We shall subdivide the circuit into LSI-size packages. First the designers effect a "natural" subdivision of the total design into "functional" boxes. Then there is SEGMENTATION, a segment being the set of all circuits feeding a given output or register and only those - registers included in segments are treated as (pseudo-)inputs. Finally these naturally subdivided segmentations are initially partitioned on a level-by-level basis and cuts inserted on the basis of minimizing these cuts - each level has its "cut" number and the level, leaving the segments more or less equal, is chosen. Cf. 1981 for further details. Wherever a cut is made a register controlling the cut is inserted, as shown in Fig. 2

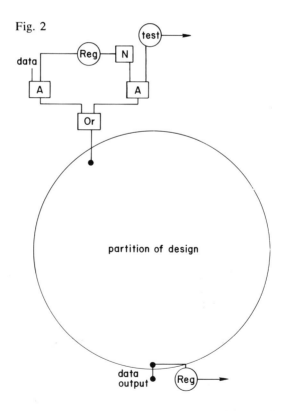

Each register for each such cut is connected at test time into a shift-register chain, emanating from a single output pin. When the register for a given cut is 1, then normal discourse takes place; when it is 0, then the test input, also connected from the exterior, is fed in. Likewise the values at the registers and the cuts are read out.

This means that the diagnosis of a VLSI design is reduced to several LSI diagnosis - and test-generation problems.

REFERENCES.

1980. Roth, J. Paul, "Computer Logic, Testing and Verification", *COMPUTER SCIENCE PRESS, Rockville, MD.*

1981. Kurtzberg, J. M., and J. P. Roth, "Reducing VLSI Diagnosis to LSI Diagnosis", IBM Technical Disclosure Bulletin, to appear.

AUTOMATING THE DESIGN OF TESTABLE HARDWARE

A.C. Parker[*] and L.J. Hafer[†]

[*]*Department of Electrical Engineering
University of Southern California
Los Angeles, CA 90007 USA*

[†]*Department of Electrical Engineering
Carnegie-Mellon University
Pittsburgh, PA 15213 USA*

1. INTRODUCTION

This paper proposes a formal method for the design of testable hardware. The method uses an algebraic model of resister-transfer (RT) behavior which has been developed (Hafer & Parker, 1981). This model has been used to synthesize register-transfer data path hardware from abstract descriptions of required function (data flow). The technique proposed here uses Hafer's modeling technique along with additional testability constraints and rules for computation of testability to produce testable register-transfer hardware.

Definitions of testability and what testable hardware means vary. In this paper, we refer to testability as a combination of six measures which are related to how difficult it is to control or observe a "zero" or "one" value at a given point in the hardware. When we state that hardware is testable, we mean that the testability measures are below some upper bound.

The research described here has three phases. The first phase is the production of a set of rules which describe computation of testability of individual register-transfer elements. The second phase involves the application of these rules to example register-transfer circuits to illustrate design techniques which increase testability. These rules are actually applied by creating graph models of the circuit controllability and observability aspects, and then solving these graphs using a complex sort of loop analysis to determine testability (Parker, 1981). The third phase involves use of the above rules, Hafer's formal model of RT behavior, and a set of constraints on testability of the resultant hardware as inputs to a synthesis program. This constitutes the entire set of constraints the hardware must meet, and which an automatic synthesis program must consider.

2. RULES FOR COMPUTATION OF TESTABILITY

The testability rules are partitioned into controllability and observability rules (there are a pair of rules for 0 and 1 controllability), according to the method specified by Goldstein and Thigpen (1980). This method is based on earlier work by Breuer and Friedman (1979). Similar research has been performed by others, the most relevant of which is the register-transfer work by Stephenson and Grason (1976). These rules are further subdivided by Goldstein into sequential and combinational rules. Thus, for each register-transfer element there are six testability rules.

Each testability rule is based loosely on input/output relationships of each RT element, and is either a linear function or involves the computation of the minimum of two or more values (which can be linearized). These rules have been derived for combinational elements (e.g., multiplexers), registers, and simple counters. (It should be noted that shift registers and counters have more complex relationships since they involve feedback loops.)

An example testability rule is

$$CC^1(Q) = CC^1(D) + CC^1(C) + CC^0(C) + CC^0(R) \tag{1}$$

which states that the ability to control the outputs (Q) of a register built of D flip flops to 1's ($CC^1(Q)$, combinational controllability) depends on the ability to control the inputs (D) to 1, to keep the CLEAR (R) input unasserted, and to cause the clock (C) to transit from 0 to 1 by controlling it to both 0 and 1. Once this computation has been done for each element, each element can be related to controllability and observability subgraphs which represent the testability rules for the element. Then, testability of the entire circuit can be computed using the graph produced by composition of the testability subgraphs of the individual elements. An example graph is shown in Fig. 1. This graph represents the controllability subgraph for the D-FF register.

The ability to express the testability relationships in a hierarchical fashion allows testability analysis to be done module by module. Furthermore, analysis of the controllability and observability graphs can be done symbolically instead of iteratively, as done by Goldstein. (This is analogous to the use of symbolic simulation as opposed to conventional simulation.)

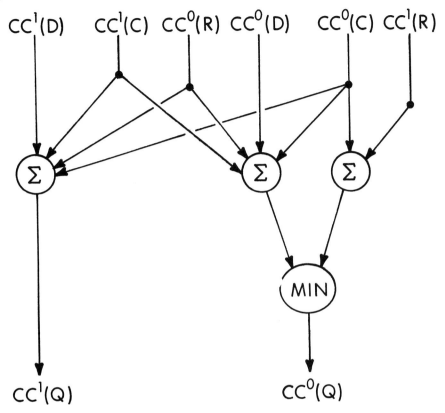

Fig. 1. *The subgraph for the D-flip-flop register.*

3. APPLICATION OF THE TESTABILITY RULES

When the above rules for testability computation are applied, designs which are produced tend to exhibit increased testability (lower controllability values) when there are shorter paths from input to output, particularly when there are feedback loops. A simple example with two input variables, four operations and two output variables illustrates varying controllability of the outputs when different RT-level implementations are examined.

4. SYNTHESIS OF TESTABLE CIRCUITS

The remainder of this paper presents a technique for automatically producing testable designs from a set of constraints on testability and function. A design is specified in data flow graph form, indicating the functional requirements of the hardware but not the register-transfer structure. A set of constraints is required to ensure proper behavior (including timing) of the resultant design. These constraints are generated automatically from the data-flow graph (see Hafer & Parker (1981)). Rules for computing testability for each variable (value) in the flow graph must also be presented. Finally, a set of constraints on the testability of the resultant design must also be given.

Simultaneous solution of the above constraints would produce designs which were not only correct and met timing constraints, but also met testability requirements. Of course, such a solution technique is computationally infeasible for large designs, but interactive, iterative design is promising. Also, the formal specification of constraints itself is important because it specifies the design problem clearly and provides a driving force for heuristic synthesis programs. We now present a simple example data-flow graph, and example rules and constraints for synthesis program input which relate to the testability and proper behavior of the synthesized design.

(a) *An Example of Design Behavior*

An example of the data-flow behavior of a design is shown in Fig. 2. This graph represents the function

$$o_{2,1} = x_2[i_{2,1}, x_1(i_{1,1}, i_{1,2})] \qquad (2)$$

where $o_{2,1}$ is the output of the design, $i_{1,1}$, $i_{1,2}$ and $i_{2,1}$ are the inputs, and x_1 and x_2 the operations performed on the inputs. Depending on speed, cost and external interface constraints, the design could be implemented with up to four registers and two functional units.

(b) *A Model of Register-transfer Behavior*

The formal model of RT-level behavior includes constraints like:

$$T_{OS}(o_{a,c}) \geqq T_{OA}(o_{a,c}) + D_{SS}(s_e) \qquad (3)$$

which states that the time an output $o_{a,c}$ is stored (T_{OS}) must be greater than or equal to the time the output becomes available (T_{OA}) plus the set-up time D_{SS} of the register S_e.

If the model is to be used for synthesis, of course, zero-one (binary) variables have to be added to cover all possible selections of registers. For the examples given in Fig. 2,

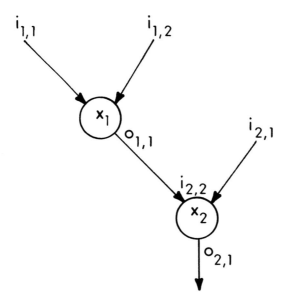

Fig. 2. *An example of data-flow behavior*

the output from x_1, $o_{1,1}$, may or may not be stored in a register. This choice can be indicated by the 0-1 variable $\gamma_{1,1}$. Thus, the general rule given in Equation 3 becomes

$$T_{OS}(o_{a,c}) \geqq \gamma_{a,c}[T_{OA}(o_a) + D_{SS}(s_e)] \qquad (4)$$

for any output $o_{a,c}$. If $\gamma_{a,c} = 0$, then the rule is ignored.

(c) *Rules for Computation of Testability*

In reality, the rules for testability conputation are much more complex than those presented in Section 2 since we refer to the controllability of variables rather than physical interconnections, which do not exist in the model. Therefore, testability rules like Equation 1 must be expanded to express the controllability of inputs and outputs of the x_a operations. For example, the combinational controllability to 1 of a given input $i_{a,b}$ would depend on where the input came from (whether it was a stored output from a function or was accessed directly from the function outputs). Thus

$$CC^1(i_{a_1,b}) = \delta_{a_1,b} \left\{ \sum_e \rho_{e,a_2,c} \left[\min_{\forall a} \rho_{e,a,c} (CC^1(o_{a_2,c})) \right. \right.$$
$$\left. + CC^1(C) + CC^0(C) + CC^0(R) \right] \Big\}$$
$$+ \omega_{a_1,b} \left\{ \sum_d \sigma_{d,a_2} \left[\min_{\forall a} \sigma_{d,a} (CC^1(o_{a_2,c})) \right] \right\} \quad (5)$$

The output value $o_{a_1,c}$ is to be used as input value $i_{a_1,b}$. $\delta_{a_1,b}$ indicates input $i_{a_1,b}$ has been accessed from stored output $o_{a_2,c}$. $\omega_{a_1,b}$ indicates it is accessed directly from output $o_{a_2,c}$ of the functional unit. The summations over the e and d actually just select the storage element or operator where $o_{a_2,c}$ is to be accessed. $\rho_{e,a,c} = 1$ indicates storage element s_e stores output $o_{a,c}$. $\sigma_{d,a}$ indicates functional unit f_d is used to implement operation x_a. Because other outputs (perhaps more controllable) can also be stored in element s_e or accessed from operator f_d, we take the minimum controllability over all possible outputs which are stored in the same s_e or accessed from the same f_d. We can write a similar rule for $CC^1(o_{a,c})$ as a function of $CC^1(i_{a,b})$ and $CC^0(i_{a,b})$. However, this rule becomes complex since each output controllability rule differs as the function performed on the inputs changes.

(d) *Testability Constraints*

In addition to specifying rules for computation of testability, the designer should also specify upper bounds on the controllability and observability values permitted in the design. An example controllability constraint is

$$\min_{\forall a} \left[\rho_{e,a,c} [CC^1(o_{a,c}) + CC^1(C) + CC^0(C) + CC^0(R)] \right.$$
$$\left. + (1-\rho_{e,a,c}) \cdot CC^1_{max} \right] \leq CC^1_{max} \quad (6)$$

for any storage element s_e. This constraint looks for the presencr of a non-zero ρ, a 0-1 variable which indicates that the output c from operation a is stored in register s_e. If no variables are stored in s_e, this value is CC^1_{max}, otherwise it is the minimum of the CC^1 values for all the operations which store values in s_e. Then, the controllability of the input to this register depends on the minimum controllability of all variables selected by $\rho = 1$ and stored in the register. CC_{max}^1 is the maximum controllability allowable at the inputs to this register. Thus, the relationship states that the minimum controllability to 1 of a set of variables which all have been put into the same register is limited to a certain upper bound. This is more complex than constraining the data path hardware itself, but it is difficult to explicitly constrain controllability of data paths since the paths do not exist prior to design, and are only implicit in the resultant design unless many more constraints are introduced into the existing RT-level model.

5. CONCLUSIONS AND DISCUSSION

This paper has proposed a method which can be used to automatically synthesize digital circuits of any desired level of testability. In practice, however, the actual design process operating off the constraints presented here must be more thoroughly researched. The value of this research at this point is that it provides greater understanding of the problems of synthesizing testable designs, merely by formalizing the design rules.

6. ACKNOWLEDGEMENTS

The authors would like to acknowledge the assistance of Mel Breuer in this research. Funds for this research were provided by the U.S. Army Research Office under Grant DAAG29-80-K-0083.

REFERENCES

Breuer, M.A. and Friedman, A.D. TEST/80 - A proposal for an advanced automatic test generation system, *Proceedings AUTOTESTCON '79*, 302-312, September 1979.

Goldstein, L.H. and Thigpen, E.L. SCOAP: Sandia controllability/observability analysis program, *Proceedings of the 17th Design Automation Conference*, 190-196, June 1980.

Hafer, L. and Parker, A.C. A formal method for the specification, analysis and design of register-transfer digital logic, *Proceedings of the 18th Design Automation Conference* (to be published), June 1981.

Parker, A.C. Toward a Methodology of Solution of Testability of Register-Transfer Circuits, Technical Memo S-3, University of Southern California, March 1981.

Stephenson, J.E. and Grason, J. A testability measure for register transfer level digital circuits, *Proceedings of the 6th Fault Tolerant Computing Symposium*, 1976.